端侧 AI 模型部署入门、原理和进阶实战

On-device AI Model Deployment Introduction, Principles, and Advanced Practice

葳 葳 编著

东南大学出版社
·南京·

内容提要

本书系统地讲解各种模型在端侧平台（含嵌入式设备、移动端设备）中的工程化实践，重点讨论模型优化（包括剪枝、蒸馏和量化等）、模型轻量化设计、高性能计算、Neon 编程、ARM 处理器 OpenCV 编程、基于 TFLite 的端侧模型部署和性能优化、NPU 和 GPU 推理加速等。通过本书，读者可以理解端侧 AI 模型部署内容，包括算法及算子优化和对模型的精度、性能的评估和调优。本书能够帮助读者朋友找到正确的学习和研究方向，以及正确的工程方法。

本书可作为高等院校计算机应用、人工智能、智能感知、元宇宙工程、机器人工程、信号处理、图像处理、嵌入式开发、电子信息工程、软件工程、集成电路设计与集成系统、微电子科学与工程等专业及学科的本科生及研究生教材，也可以作为相关领域的科研和工程技术人员的参考书籍。

图书在版编目（CIP）数据

端侧 AI 模型部署入门、原理和进阶实战 / 葳葳编著．
南京：东南大学出版社，2024.11. -- ISBN 978-7-5766-1577-7

Ⅰ.TP181
中国国家版本馆 CIP 数据核字第 2024HC1780 号

责任编辑：张 煦　　责任校对：韩小亮　　封面设计：王 玥　　责任印制：周荣虎

端侧 AI 模型部署入门、原理和进阶实战
Duance AI Moxing Bushu Rumen、Yuanli He Jinjie Shizhan

编　　著	葳　葳
出版发行	东南大学出版社
社　　址	南京市四牌楼 2 号　邮编：210096　电话：025-83793330
出 版 人	白云飞
网　　址	http://www.seupress.com
经　　销	全国各地新华书店
印　　刷	广东虎彩云印刷有限公司
开　　本	787 mm × 1092 mm　1/16
印　　张	22.50
字　　数	479 千字
版　　次	2024 年 11 月第 1 版
印　　次	2024 年 11 月第 1 次印刷
书　　号	ISBN 978-7-5766-1577-7
定　　价	98.00 元

本社图书若有印装质量问题，请直接与营销部联系调换。电话（传真）：025-83791830

前　言

随着 AI 技术的不断发展以及 AI 技术在行业场景中的应用不断深入，对端侧 AI 模型部署的需求越来越多，特别是在智能驾驶、车载、辅助驾驶、安防、机器人、人工智能、数字化转型、AR/VR/XR、元宇宙、语音识别、自然语言处理、人工智能翻译等领域，急需这方面的研发和管理等人才。

目前市场上关于端侧 AI 模型部署方面的书籍并不丰富，且其中有些书籍不够系统化，比如只讲单个端侧模型部署框架，难以让读者对端侧 AI 模型部署有全面系统的了解。本书系统讨论了目前行业多种部署框架。通过本书，读者朋友可以明白如何从事端侧 AI 开发，特别是其中的算法优化部分（更具体点说是让读者明白如何去开发算法并且通过优化部署到嵌入式设备或智能手机等智能终端中去），以及精度和性能的评估和调优。

本书共 12 章，各章节的介绍如下：

第 1—3 章介绍准备工作，其中第 1 章介绍端侧 AI 的应用，算法优化，编程基础等。在第 2 章，我们将理解什么是指令集架构、处理器架构、异构计算芯片，以及端侧 AI 芯片等。在第 3 章，我们将掌握 Shell 命令解释器和脚本，环境变量配置，编译器 GCC，CMake 项目构建、make 编译，LLVM 和 Clang，Android SDK 和 NDK 配置，Clang/Clang++ 编译 C/C++/Neon 程序和终端运行，GDB 调试，ADB 工具等知识和技能。

第 4 章对目前比较主流的移动端推理框架有一个全面的介绍，对算法部署实践有直接的指导作用。

第 5 章介绍模型压缩，对于后期产品开发中的技术方案选定具有非常重要的指导作用。

第 6 章介绍端侧模型部署框架，包括 ARM ComputeLibrary/Arm NN、NNAPI、TensorRT、TNN/NCNN、OpenVINO、TFLite、Core ML、RKNN SDK、SNPE/QNN、MNN、MediaPipe、NeuroPilot 等。

第 7 章介绍 ARM 处理器 OpenCV 编程，内容包括 ARM 平台移植 OpenCV 库，OpenCV 库编译错误及解决方法，ARM 平台 C/C++ 图像处理实例等。

第 8 章介绍 Neon 指令和内联函数，内容包括 Neon 寄存器和 Neon 指令类型、Neon

编程方式和 Neon 内联函数、基于指令功能的 Neon 指令分类、Neon 支持库。通过本章可以明白如何使用 Neon 编程实现算法的优化。

第 9 章介绍基于 TFLite 的端侧模型部署和性能优化，内容包括 TFLite 委托，TFLite 交叉编译和部署——基于 ARM 平台，YOLOv8 模型端侧部署——低空无人机巡检等。

第 10 章介绍 NPU 推理加速——无人机采茶机器人研发实践，内容包括基于 ONNX 格式模型转换和验证，模型量化误差评估，NPU 推理性能和内存使用情况评估等。

第 11 章介绍端侧 GPU 硬件加速模型推理——智能水下摄像机器人研发实践，内容包括 Transformer 模型端侧部署评估。

第 12 章介绍安全智能，以隐私 OCR 为实例讨论 AI 服务的安全重要性。不能只有智能，而没有安全。

本书可作为高等院校计算机应用、软件工程、人工智能、智能科学、元宇宙工程、机器人工程、信号处理、图像处理、嵌入式开发、电子信息工程、集成电路设计与集成系统、微电子科学与工程等专业及学科的本科生及研究生教材，也可以供相关领域的科研和工程技术人员参考。"乞火不若取燧，寄汲不若凿井"[①] 希望这本书能够做到授人以渔。

本书中的参考文献均在文中的脚注中说明，便于读者朋友进一步研究和实践。

考虑到领域知识的系统性、技术的发展更新、笔者的知识局限性以及本书的篇幅，部分内容没有做深入讲述，读者朋友可以在此基础上参考相关的资料去做进一步的学习和实践。

在此感谢家人的支持和理解，感谢东南大学出版社张煦等各位编辑老师的帮助和指点。由于作者水平有限，书中难免会有疏漏及其他欠妥之处（包括文献和网络资源的引用和标注），在此恳请广大读者朋友指点、交流和反馈。

葳 葳
2024 年 7 月

[①] 出自《淮南子·览冥训》

目 录

第 1 章 端侧 AI 概述 ·· 001
 1.1 什么是端侧 AI ·· 002
 1.2 端侧模型性能优化 ··· 012
 1.3 端侧模型低功耗设计 ·· 026
 1.4 端侧模型加固 ··· 027

第 2 章 端侧 AI 芯片 ·· 029
 2.1 中央处理器体系结构 ·· 030
 2.2 指令集架构 ·· 031
 2.3 处理器架构及异构计算芯片 ··· 032
 2.4 端侧 AI 芯片 ·· 035
  2.4.1 ARM ··· 035
  2.4.2 DSP ·· 040
  2.4.3 GPU ·· 048
  2.4.4 FPGA ·· 048
  2.4.5 NPU ·· 049
  2.4.6 类脑芯片 ·· 051

第 3 章 Linux 开发环境及工具介绍 ··· 053
 3.1 Shell 命令解释器和脚本 ·· 054
 3.2 环境变量配置 ··· 056
 3.3 编译器 GCC ··· 057
 3.4 CMake 项目构建、Make 编译 ·· 058
 3.5 LLVM 和 Clang ··· 060
 3.6 Android SDK 和 NDK 配置 ·· 061
 3.7 Clang/Clang++ 编译 C/C++/Neon 程序和终端运行 ························· 063

3.8　GDB 调试 ·· 066
　　3.9　ADB 工具 ·· 068

第 4 章　算子和图优化 ··· 071
　　4.1　算法层优化 ·· 072
　　　　4.1.1　Img2col+GEMM 优化卷积 ··· 072
　　　　4.1.2　Winograd 优化卷积 ·· 078
　　4.2　硬件层优化 ·· 080
　　　　4.2.1　SIMD 指令向量化 ··· 080
　　　　4.2.2　多核 CPU 中 OpenMP 编程 ·· 083
　　　　4.2.3　GPU 并行计算 ··· 088
　　　　4.2.4　Cache 优化 ·· 099
　　4.3　图优化 ··· 105
　　4.4　AI 编译器 ··· 107

第 5 章　模型压缩 ·· 109
　　5.1　轻量化网络模型设计 ·· 110
　　　　5.1.1　下采样 ·· 110
　　　　5.1.2　上采样 ·· 116
　　　　5.1.3　全局池化 ··· 122
　　　　5.1.4　分组卷积 ··· 126
　　　　5.1.5　全局加权池化 ··· 127
　　　　5.1.6　1×1 卷积 ··· 129
　　　　5.1.7　深度卷积 ··· 134
　　　　5.1.8　逐点卷积 ··· 134
　　　　5.1.9　异构卷积 ··· 134
　　　　5.1.10　深度可分离卷积 ·· 136
　　　　5.1.11　空洞卷积 ··· 137
　　　　5.1.12　跳跃连接 ··· 138
　　　　5.1.13　Flatten ·· 140
　　　　5.1.14　BatchNormalization ·· 141
　　　　5.1.15　Dropout ·· 143
　　　　5.1.16　全连接层 ··· 144
　　　　5.1.17　SENet ··· 144

	5.1.18 MobileNet	145
	5.1.19 注意力机制	147
	5.1.20 新算子和模型的探索	150
5.2	剪枝	151
	5.2.1 结构化剪枝	151
	5.2.2 非结构化剪枝	153
5.3	网络架构搜索	154
5.4	低秩分解	155
5.5	知识蒸馏	156
5.6	量化	158
	5.6.1 量化原理	160
	5.6.2 对称量化/非对称量化	163
	5.6.3 伪量化节点	163
	5.6.4 训练后量化/量化感知训练	163
	5.6.5 量化提升策略	164
	5.6.6 量化感知训练框架介绍	166

第 6 章 端侧模型部署框架 ... 177

- 6.1 ARM ComputeLibray/Arm NN ... 178
- 6.2 NNAPI ... 179
- 6.3 TensorRT ... 180
- 6.4 TNN/NCNN ... 180
- 6.5 OpenVINO ... 181
- 6.6 TFLite ... 181
- 6.7 Core ML ... 182
- 6.8 RKNN SDK ... 182
- 6.9 SNPE/QNN ... 182
- 6.10 MNN ... 184
- 6.11 MediaPipe ... 185
- 6.12 NeuroPilot ... 185

第 7 章 ARM 处理器 OpenCV 编程 ... 187

- 7.1 ARM 平台移植 OpenCV 库 ... 188
- 7.2 OpenCV 库编译错误及解决方法 ... 193

7.3 ARM 平台 C/C++ 图像处理实例 ··· 197
7.3.1 图像浅拷贝和深拷贝 ··· 197
7.3.2 颜色空间转换 ··· 199
7.3.3 图像二值化 ··· 200
7.3.4 图像翻转 / 旋转 / 缩放 / 裁剪 ··· 203
7.3.5 二维码检测和解码 ··· 205

第 8 章 Neon 指令集加速算法和算子底层指令加速 ··· 209
8.1 Neon 寄存器和数据类型 ··· 210
8.2 Neon 指令类型 ··· 213
8.3 Neon 编程方式和内联函数 ··· 214
8.3.1 Neon 汇编指令 ··· 214
8.3.2 编译器自动向量化 ··· 214
8.3.3 Neon 第三方库 ··· 215
8.3.4 Neon 内联函数 ··· 219
8.4 Neon 常用内联函数介绍 ··· 226
8.4.1 类型转换指令 ··· 226
8.4.2 加载存储指令 ··· 228
8.4.3 算术运算指令 ··· 240
8.4.4 数据处理指令 ··· 242
8.4.5 向量乘法指令 ··· 245
8.4.6 逻辑和比较运算指令 ··· 245
8.4.7 浮点指令 ··· 245
8.4.8 移位指令 ··· 252
8.4.9 置换指令 ··· 252
8.4.10 其他指令 ··· 261
8.5 Neon 编程优化算法实例 ··· 263
8.5.1 RGB 转 Gray 颜色空间 ··· 263
8.5.2 RGB 内存空间解交织和交织 ··· 265
8.5.3 矩阵乘法性能优化 ··· 269
8.6 NEON2SSE 介绍 ··· 272

第 9 章 基于 TFLite 的端侧模型部署和性能优化 ··· 273
9.1 TFLite 委托 ··· 274

9.2 TFLite 交叉编译和部署——基于 ARM 平台 ·········· 274
    9.2.1 用 CMake 工具构建 TFLite ·········· 275
    9.2.2 XNNPACK 编译问题定位和解决 ·········· 279
9.3 YOLOv8 模型端侧部署——低空无人机巡检 ·········· 282
    9.3.1 TFLite FlatBuffer 模型文件转换和验证 ·········· 283
    9.3.2 ARM 平台模型部署——C 语言版 ·········· 287
    9.3.3 部署性能优化 ·········· 298

第 10 章 NPU 推理加速——无人机采茶机器人研发实践 ·········· 299
10.1 基于 ONNX 格式模型转换和验证 ·········· 302
10.2 模型量化误差评估 ·········· 308
10.3 NPU 推理性能和内存使用情况评估 ·········· 310

第 11 章 端侧 GPU 硬件加速模型推理——智能水下摄像机器人研发实践 ·········· 313
11.1 水下图像增强模型——Transformer 模型端侧部署评估 ·········· 314
    11.1.1 水下图像增强模型网络结构和可视化分析 ·········· 315
    11.1.2 基于 QNN 推理框架 GPU 核部署评估 ·········· 318
11.2 水下目标检测模型——YOLOv8 模型端侧部署评估 ·········· 321

第 12 章 安全智能——以隐私 OCR 为实例 ·········· 325
12.1 OCR 的四个矛盾 ·········· 326
12.2 OCR 模块 ·········· 331
    12.2.1 识别主体区域检测 ·········· 331
    12.2.2 文本定位 ·········· 331
    12.2.3 字符识别 ·········· 332
12.3 隐私 OCR ·········· 333
    12.3.1 隐私 OCR 流程 ·········· 333
    12.3.2 端侧敏感信息智能脱敏方法 ·········· 334

参考文献 ·········· 336
后　　记 ·········· 346

第 1 章

端侧 AI 概述

人工智能（Artificial Intelligence，AI）的出现，就像一场春雨后，满山的春笋都在争先恐后般破土而出，发出"沙沙"的声音。AI和蒸汽机、电力、信息技术一样对人类社会的发展产生巨大、深远和广泛的影响，在不断赋能各行各业，对社会发展具有颠覆性的推动力。

1.1 什么是端侧 AI

AI 研究方向

AI 的研究方向包括算法、框架、软件、芯片和场景等（如图 1-1 所示），具体可以分为以下几个部分。

图 1-1　AI 研究领域全景图

（1）AI 基础理论

AI 基础理论包括数学、计算机科学、信息论、控制论等。其中数学更是 AI 的基石，涉及线性代数（如引入向量和矩阵用于数据表示和计算）、微积分（如梯度下降法和反向传播算法）、概率论与数理统计等。

（2）AI 芯片及架构

近年芯片以及其中的 AI 芯片是行业热门领域，芯片和算法一样作为 AI 行业的基石，芯片中算法落地场景更是"如火如荼"，它对我们的工程人员提出了更高的要求，不仅要精通算法，还要精通硬件，能够实现模型裁剪、量化以及芯片指令集层次优化算法，把控算法的精度、处理性能和功耗指标。正如《离骚》有云"路漫漫其修远兮，吾将上下而求索"，这需要不断研究和创新，不断提升专业分析和解决问题的能力，不断做好服务部署以及服务迭代升级。人工智能行业中有个岗位叫 AI 系统/硬件架构师（涉及 AI 平台架构设计、NPU 架构开发、ASIC 设计等），读者朋友是不是很感兴趣呢？

(3)平台、框架和部署

深度学习框架是支持开发和训练深度学习模型的整体解决方案。深度学习框架的能力体现在训练模型的能力、部署的能力等。关于训练模型的能力，不同框架中的关键组件优化器[①]的能力有差异，不同的框架对同样的模型，其收敛的程度往往有差距，特别是针对复杂的模型任务。深度学习框架有 PyTorch、TensorFlow、Keras、MXNet、PaddlePaddle（飞桨）等。对于这些机器学习框架，在 AI 产品开发中应该如何选择？关于这个问题，笔者认为模型的开发效率也是其中一个重要的考量，它涉及现状基础、已掌握技能、学习成本、团队协作，还有成功参考实例等。选择擅长的一个框架去预研模型，等模型预研完成后，再评估选择适合的训练框架去训练量产模型。更重要的一点：开发过程是"山重水复疑无路，柳暗花明又一村"，往往到一个阶段，伴随着框架和工具的升级，之前的难点和阻碍也就烟消云散，但又遇到新的问题，就需要不断升级打怪。举个实际中出现的例子：在模型推理中，起初这个移动端推理框架不支持某一个算子，自定义算子效果又不理想；后来这个框架升级，新增支持这个算子，这一下子让我们轻松了很多。

(4)基础软件

比如模型网络结构可视化工具、模型量化中量化参数分布统计工具、模型部署中模型加密工具、模型训练中的数据预标注工具等等。

(5)算法和模型

涉及大模型分布式训练和推理、超大规模预训练模型、文本生成、多模态，轻量化模型应用，异构计算。

(6)伦理和政策

随着 AI 的发展，特别是人形机器人、大模型的发展，AI 伦理和治理越来越重要。人类社会既要智能，也要安全。AI 向善是人类共同的愿景。

(7)AI 产品化

AI 产品化的首要任务是什么？笔者认为是做好产品定义。产品定义（Product definition）的内容包括产品愿景、市场定位、产品形态和功能设计、产品版本规划（短期目标和长期目标）等。产品定义不能简单拍脑袋决定，产品目标切不可摇摆不定，唯有做到"如切如磋，如琢如磨"这八个字，对场景、需求、痛点做深入评估，思考如何用好资源，做到既有效率又有质量。产品定义不等于产品开发，要有更开阔的视野和更大的格局。我们容易从 0 做到 1，但往往很难从 80 做到 90。产品定义需要用心，从用户角度去深入思考。对于开发一款 AI 产品，我们需要定义这款产品的功能，思考这些功能的合理性、实用性和市场潜力，评估目前的资源能否支持开发以及在开发中涉及的方案调整。这样让算法优化有更明确的方向：是打造产品，打磨产品！而不是空谈技术！

---

① 调整模型参数，使损失函数最小化。如何高效调整参数，体现的是框架的能力

端侧 AI

AI 模型部署的硬件平台有端、边、云。我们列举一些端侧模型应用的场景。

场景 1：智能手机中的 AI 算法。

（1）拍照模式中人脸检测和人脸识别。

（2）AI 消除功能如图 1-2 所示，借助手机端侧的算力，用户可以很轻松地消除图片中不需要的字符、路人或者某个物体，从中间的子图，可以看到智能消除应用到了实例分割的模型，而且消除后的图片让用户感觉不到丝毫的涂抹痕迹，从而提高用户的拍照体验。

图 1-2　AI 消除图片中的物体

（3）文档矫正功能（如图 1-3 所示）。

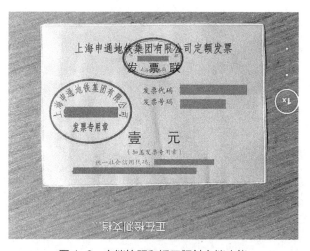

图 1-3　文档拍照和矫正倾斜文档功能

关于场景 1，笔者一直秉持产品重在用户体验的观点。AI 技术在不断发展，AI 产品也像雨后的春笋一个接一个冒尖，但其中不少"想法很丰满，现实很骨感"，犹如鸡肋，食之无味，弃之却又可惜。我们大声呼喊：如何打造一款好的 AI 产品呢？笔者认为，一个优秀的工程师也必须是一个优秀的产品体验官——代码需要有灵魂，不是机械地实现。产品体验官也不是那么容易当的，它重在体验、思考和洞察，需要有敏锐观察力和细微感受力。我们需要对产品进行深入熟悉和理解，做到内心期待认可，"不用心去烹饪，很难做出一道美食"。所以产品体验官是感性的，更是理性的。

（4）人脸关键点检测。

人脸关键点检测（Face Landmark Detection）或叫人脸对齐（Face Alignment）是实现精确定位标记人脸的脸颊、眉毛、眼睛、卧蚕、鼻子、嘴巴等和人脸轮廓的关键点位置的算法。开源的模型很多，如 Dlib、MediaPipe，其中 Dlib 提供的 shape_predictor_68_face_landmarks.dat 参数模型可帮助我们快速实现基本功能，如图 1-4 所示是人脸关键点检测，如图 1-5 所示是嘴巴关键点检测。[①] 如果使用 MediaPipe，不仅可实现人脸 468 个关键点的定位，而且精度更高，还可以实现人体姿态检测，想想功能有多么强大。

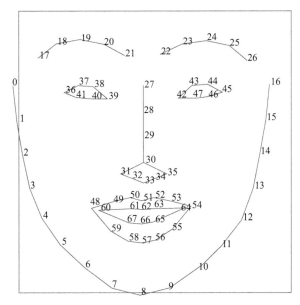

图 1-4　人脸检测和人脸关键点检测

图 1-5　嘴巴关键点检测

基于人脸关键点检测，可以实现人脸识别、活体检测、反深度伪造、人脸搜索（实现手机相册中同一人物的检索）、人脸网格、眨眼检测、脸部特效、口型/动作生成、3D 人脸重建、人脸属性识别、人脸肤质评估等功能。它在视频交互、AR/VR 交互、数字人、短视频特效、自拍分割、视频直播等领域具有很大的想象空间。

---

① 官网地址：http://dlib.net/

场景2：无人机在勘察、航拍、搜救、农业喷洒、运输，以及无人机灯光秀等场景中，缺少不了AI算法。

（1）手势识别应用到无人机交互中。通过OK和胜利等手势控制无人机的拍照和录像功能，通过手掌的移动控制无人机的上下左右运动，如图1-6所示。

图1-6　用手势识别控制无人机

（2）无人机图传技术。无人机搭载的相机拍摄的视频通过视频编解码、无线传输、抗干扰等技术高清流畅且低延时传输到接收设备。

场景3：智能终端具有AI属性。

（1）智能音箱。植入了智能语音交互技术，赋予了这款音箱AI的属性（如图1-7所示）。

图1-7　Amazon推出的Echo智能音箱[①]

---

① 图来源：https://www.amazon.com

（2）智能手环。智能手环的健康监测功能，包括心率、体温、血氧饱和度等指标监测，运动评估，久坐提醒，睡眠监测，以及无创血糖监测等。

场景4：辅助驾驶应用AI功能。

（1）辅助驾驶中注意力集中监测功能。

（2）倒车影像中倒车辅助线功能。

（3）开门防后撞影像。

（4）泊车系统专家，如图1-8所示。

图1-8　泊车系统专家示意图①

（5）环车安全影像、流媒体后视镜。

（6）AR-HUD（HUD全称为Head Up Display，平视显示器）。

4.7 拾音器中自适应去噪、均衡和去回音模块。

场景5：机器人。

（1）机器人是人工智能的风向标。波士顿仿人机器人Atlas让我们大开眼界（如图1-9（a）所示）。工业机器人带来指数级生产力。家用机器人作为智能家居的新宠，在逐渐改善我们的生活：可以给我们做一道美味佳肴，相信未来可以辅助我们照顾老人和照看小孩。机器人！机器人！机器人！前景很好！

---

① 图来源：https://baike.baidu.com/item/%E8%87%AA%E5%8A%A8%E6%B3%8A%E8%BD%A6/8187028?fr=ge_ala

（a）波士顿仿人机器人[1]　　　　　　　　（b）机械臂[2]

图1-9　机器人

（2）酒店配送机器人几乎是现代酒店的标配。它可以自主乘电梯、避障（结合内置超声波雷达、深度摄像头、激光雷达等多传感器信息融合定位实现）、智能拨打电话、自动回充、智能创建地图等。

场景6：时间序列异常检测。

（1）心电图（Electrocardiogram，ECG）分析。

智能手表和手环内置的人体心率检测和心电图功能，对用户的健康和生活方式起到一定的关怀指导作用。

（2）机械振动信号异常检测。

通过对机械振动信号的检测，结合历史的数据，可分析预判设备的潜在隐患，从而避免故障的发生。比如检测新能源电动车电机的声音、电池充电的声音，避免事故的发生。

（3）水质异常监测、生态环境监测。如图1-10所示，通过在容易产生垃圾的区域或者需要重点监测的区域安装监控摄像头来实现水域异常状况检测。通过无人机航拍对河道的巡检，实现河道中水面垃圾区域检测、排污口检测、水体污染检测、水的颜色异常检测等。

图1-10　AI水质监测

---

[1] 图来源：https://www.bostondynamics.com/
[2] 图来源：https://blog.robotiq.com/whats-new-in-robotics-015.03.2024

场景 7：AIGC（AI-Generated Content），即生成式 AI 大模型，给人工智能带来了更大的想象空间，端云结合，AI 未来可期。

（1）ChatGPT（Generative Pre-trained Transformer，生成型预训练变换模型），如图 1-11 所示为大语言模型发展概况。

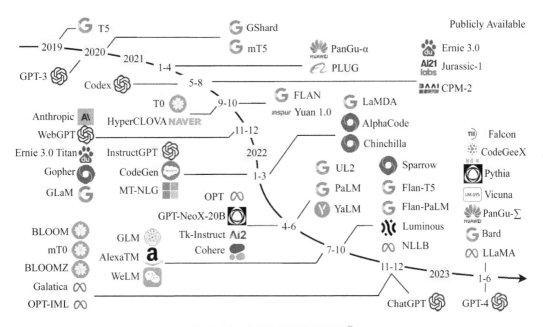

图 1-11 大语言模型发展概况[①]

（2）Sora 为文生视频大模型，它可以根据输入的一段描述视频场景的提示词，生成一段视频，视频让人眼前一亮，如图 1-12 所示。

图 1-12 Sora 生成 wooly-mammoth 视频某一帧[②]

（3）Stable Diffusion 和 Midjourney 是 AI 绘画生成工具，目前前者开源而后者不开源。图 1-13 所示为 Midjourney 生成图。

---

① 图来源：A Survey of Large Language Models，https://arxiv.org/pdf/2303.18223.pdf
② 图来源：https://cdn.openai.com/sora/videos/wooly-mammoth.mp4

图 1-13　Midjourney 生成图

场景 8：智能安防通过 AI 视频分析技术的加持，让安防应用场景变得更加细分。

（1）运动区域检测，也叫移动侦测，如图 1-14 所示为背景相减方法实现的连续视频帧中运动区域检测。

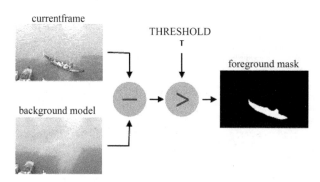

图 1-14　背景相减方法（Background subtraction）[①]

（2）越界检测。

越界检测是通过计算机视觉识别算法对安防摄像头的采集画面中设定区域实现人员越界检测和报警，从而实现安全预警和提高人工监管效率。其应用场景有水库危险区域、高楼天台限制区域、火车铁轨区域、生产机器作业危险区域等。

（3）遗留物检测。

遗留物检测能实现在车站、机场、酒店、银行、商场、电影院等公共场所中失主遗留的物品检测。

（4）车牌识别。

车牌识别是通过计算机视觉识别算法实现车辆的车牌定位、车牌颜色识别、车牌字

---

① 图来源：https://docs.opencab.org/4.x/d1/dc5/tutorial_background_subtraction.html

符信息识别和车牌异常检测（包括车牌清晰度、完整性、是否遮挡和污损、车牌真假等）等。所遇到的挑战有强光、光照不足、逆光、恶劣天气、车辆高速行驶等。车牌识别可实现公路卡口、园区/小区、停车场等场景的智能化管理。

（5）智能门锁。

智能门锁融合人脸识别（如3D人脸识别技术）、指纹、密码、NFC（Near Field Communication，近场通信）感应、高清猫眼等一系列智能技术，赋予钥匙多种形态，让锁更智能。

（6）门禁闸机。

门禁闸机在出入车站、机场、地铁、写字楼、图书馆等地方时经常见到。通过刷脸等技术的加持，让门禁闸机更加便利。

端侧模型优势和工程化挑战

以上场景中，有些场景如果采用云端的部署方案，由于网络通信质量和时延，算法实时性可能很难保证。比如手机拍照场景，模型如果部署在云端，对用户图片隐私安全也很难保证。如果算法部署在本地，也就是端侧，就能很好地解决这个问题，还可以减轻服务器负担和降低成本。

当然端侧模型的部署也就是端侧模型的工程化所面临的技术挑战也很大，如芯片算力约束、芯片功耗、发热、续航等问题，如表1-1所示。

我们可以预测：端侧算力将越来越优，端侧AI，具体到端侧模型，包括端侧大模型，在隐私保护、响应时延、功耗等方面的优势会不断发挥，它将和边缘及云端AI相辅相成一起服务社会。对，是相辅相成！

表1-1 端侧模型优势和工程化挑战

优势	挑战
实时，对处理来自摄像头的连续帧的视频应用至关重要	推理性能（推理速度、内存限制、功耗耗电量限制）；算法、指令、访存复杂度，实际计算量
离线服务，可以不依赖网络，提高用户体验	模型体积（模型在硬件设备上所占用的存储空间）的硬件制约
保护隐私	涉及模型安全问题，需要加密防破解
节省带宽和算力	平台适配
精度	模型量化，量化精度丢失

通过以上内容，我们已经明白什么是端侧AI以及端侧模型优势和工程化挑战。本书将重点讨论如何实现端侧模型的工程化，包括精度、平台、性能、功耗等。

1.2 端侧模型性能优化

先列举几个例子来讨论端侧模型性能优化。

例子 1：如图 1-15（a）所示，用户需要使用表格识别功能，但是加载的图片需要用户用手拖动定位到具体的表格边框区域，如图 1-11（b）(c) 所示，如何让软件自动定位呢？这就是算法需要去实现的功能，有这个功能，用户的体验感就更好。那么模型优化需要做什么呢？首先需要把算法的精度优化得更好，比如表格的定位更准；然后把模型压缩到一定的体积，比如 5M 大小，把推理时间优化到 10ms 以内，还要保证算法的效果基本没有受到影响。只有这样，用户在手机上使用这个功能的体验感才会非常好。

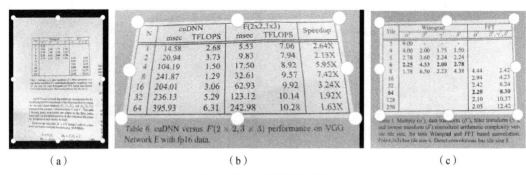

图 1-15　表格扫描和定位

例子 2：图 1-16 是证件打印机打印出来的证件效果，您是不是发现表格和字段有比较明显的错位呢，其实这是一个非常普遍的问题。在各种发票打印中，也经常出现。如何用算法去改进我们的打印机呢？我们的算法工程师提出了很好的解决方案，那就是用计算机视觉检测到正确的位置，然后控制打印机的打印位置，是不是很妙！在这里面您需要考虑一下打印的速度和算法的性能所带来的制约，这就是算法优化的任务。

课程名称	端侧AI：算法优化教程

图 1-16　证件打印机打印效果

例子 3：自动驾驶。

自动驾驶的系统包括感知、决策、控制三个部分。其中感知部分涉及多种类型的传感器和感知算法。

传感器有激光雷达、毫米波雷达、超声波雷达、车载摄像头、IMU（Inertial Measurement

Unit，惯性传感器）、GPS、里程表等。激光雷达具有测量精度高、抗干扰能力低的特点。毫米波雷达采用频率为 30～300 GHz（波长为 1～10 mm）的毫米波。毫米波波段的电磁波，可穿透烟、雾、灰尘。超声波雷达发射出的超声波对于任何材质障碍物均可反射，在短距离范围里具有测量精度高的优点，因此非常适合倒车辅助和自动泊车场景。车载摄像头属于车规产品范畴，它比普通的消费电子产品要求更高，如能够适应剧烈温差变化，抗震动，防电磁脉冲，防水，使用寿命一般要求 10 年以上，高信噪比，宽动态（适应光线剧烈变化），LED 频闪抑制（LFM），以及视场角（Field of View，FOV）、分辨率、帧率、有效探测距离等参数均需要满足实际要求。车载摄像头（包括红外摄像头）根据安装在车子上的位置，可分为前视摄像头、后视摄像头、侧视摄像头、内置摄像头和 360° 环视（前后两侧等摄像头影像拼接得到全景图）等，如图 1-17 所示。

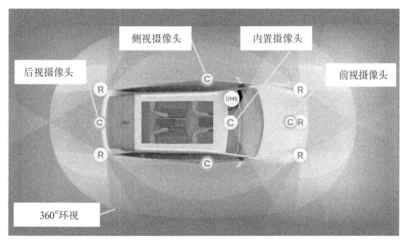

图 1-17　车载摄像头示意图[①]

感知算法有语义分割、2D/3D 目标检测（如障碍物检测、车辆/行人/交通标志检测、车轮检测）、目标跟踪、车道线/路沿检测、行为分析等，其中的模型一般为多任务检测模型。目前的感知算法方案主要分为纯视觉感知和多模态融合感知，前者以特斯拉开发的 Occupancy Networks（占用网络）神经网络为代表，后者的实现为将 Camera/LiDAR/Radar 等多个传感器采集的数据按照一定的方法处理后输出结果。该方法一般分为前融合与后融合，其中前融合先把数据按照一定的方法融合，再输入感知模块中，而后融合则是先对每个传感器采集的数据单独进行感知，再把这些感知输出通过算法融合后输出综合结果，该方法称为多传感器信息融合（Multi-Sensor Information Fusion，MSIF）。以上多传感器信息融合中还会用到诸如 Kalman Filter（卡尔曼滤波）、EKF（扩

---

① 图来源：https://www.qualcomm.com/products/automotive/automated-driving（有改动）

展卡尔曼滤波）、UKF（无迹卡尔曼滤波）、Particle Filter（粒子滤波）、Kalman Smoother（卡尔曼平滑）等滤波算法。

点云（Point Cloud）是指描述 3D 视觉世界的数据集合，其中每一个点包含的信息有 X、Y、Z 三维坐标几何位置、颜色信息或强度信息。点云除了应用在无人驾驶领域，在元宇宙、AR/VR/XR、机器人、3D 视觉、无人机等场景也具有很重要的用途。

Open3D 是一个支持点云数据处理软件快速开发的开源库，如图 1-18 所示是 Open3D 演示的效果。

图 1-18　Open3D 演示效果

以下是安装和验证 Python 版本 Open3D 的方法。

```
pip install open3d-python
>>> from open3d import *
>>>
```

以下是 Open3D 简单的测试代码：

```
from open3d import *
import numpy as np
points_ =[[ 0.00,  0.00,  0.00 ],
         [ 0.00,  0.00, 10.00 ],
         [ 0.00,  0.00, 20.00 ],
         [ 0.00,  0.00, 30.00 ],
         [ 0.00, 10.00,  0.00 ],
         [ 0.00, 20.00,  0.00 ],
         [ 0.00, 30.00,  0.00 ],
         [10.00,  0.00,  0.00 ],
         [20.00,  0.00,  0.00 ],
         [30.00,  0.00,  0.00 ]]
```

points = np.array（points_）
point_cloud = PointCloud（）
point_cloud.points = Vector3dVector（points）
draw_geometries（[point_cloud]）

其中 Open3D 过程调试如图 1-19（a）所示，Open3D 测试效果如图 1-19（b）所示，图中的点代表点云的分布图，可以看到 10 个点组成 X、Y、Z 坐标的形状。

 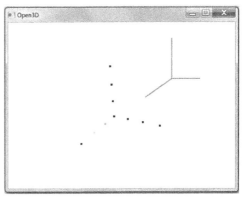

（a）Open3D 过程调试　　　　　　　（b）Open3D 测试效果

图 1-19　Open 3D 过程调试和测试效果

例子 4：ADAS。

ADAS（Advanced Driver Assistance Systems 先进驾驶辅助系统），和自动驾驶一样依赖传感器和算法实现感知、决策和控制，可以看成自动驾驶的低级阶段。其功能有很多，比如驾驶员疲劳探测（Driver Drowsiness Detection，DDD）、行人保护系统（Pedestrian Protection System，PPS）、车道保持系统（Lane Keeping System，LKS）、车道偏离预警系统（Lane Departure Warning System，LDWS）、前碰撞预防系统（Forward Collision Warning，FCW）、盲点探测系统（Blind Spot Detection，BSD）、夜视系统（Night Vision System，NVS）、自适应灯光控制（Adaptive Light Control，ALC）、自动泊车辅助（Automatic Parking Assist，APA）、交通标志识别（Traffic Sign Recognition，TSR）、陡坡缓降控制系统（Hill Descent Control，HDC）等。

例子 5：智能座舱。

智能座舱（Intelligent Cabin）是对传统汽车驾驶舱的升级，借助高性能车载芯片、人工智能等技术构建车内一体化数字平台，进一步提升用户驾乘舒适性和行驶安全性。它包括高性能车载芯片、车载大屏等硬件，智能感知服务、智能语音交互、聊天机器人（类似 ChatGPT）等软件。

例子 6：智能汽车底盘。

汽车底盘系统是自动驾驶的执行机构，是汽车的核心部件，是汽车驾乘舒适性和安全性的重要基础。智能底盘让车的底盘变得聪明起来，如主动稳定杆、智能制动系统、主动悬架（如双目立体视觉"魔毯"功能）等，这将是车企新的发力点。

例子7：低空经济——无人机载合成孔径雷达（SAR）技术应用。

低空经济是以各种有人驾驶和无人驾驶航空器的各类低空飞行活动为牵引，辐射带动相关领域融合发展的综合性经济形态。[①] 无人机是低空经济的主导产业，具有广阔的行业应用前景，如这里讨论的无人机载SAR（Synthetic Aperture Radar，合成孔径雷达）。无人机充当SAR的载荷，能发挥其灵活和高效的优势，让SAR可以迅速采集多区域多检测点多角度的高质量的数据。

SAR属于有源主动式微波成像遥感技术。雷达的孔径有限，SAR通过雷达的运动合成一个虚拟的孔径雷达天线，也就是如图1-20所示的合成孔径，从而提高SAR成像的分辨率。

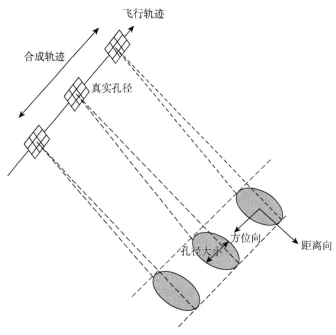

图1-20　合成孔径雷达示意图[②]

SAR图像对比光学图像的优势有：SAR具有穿透力强的特性，能够穿透云层、雾、霾、烟、灰尘、雨雪等，以及地表覆盖物（如植被、土壤等）。它不受天气和光照影响，实现全天候（没有昼夜区分）工作。这些特性是光学图像所不具备的。虽然，SAR图像

---

① 摘自百度百科，https://baike.baidu.com/item/%E4%BD%8E%E7%A9%BA%E7%BB%8F%E6%B5%8E/50884294?fr=ge_ala

② 图参考来源：https://www.jl1mall.com/forum/PostDetail?postId=20230627141600033141 8

的空间分辨率比不上光学图像,它没有颜色信息和清晰的纹理信息。但是 SAR 具有高时域分辨率特性,可用于检测运动和变化。

举例说明一下:如图 1-21 所示,由于云层的遮挡,可见光卫星往往无法获取云层下的地面洪水信息,通过 SAR 可以实现洪水监测。雷达影像中黑色区域为水体覆盖区域,图中白色虚线圈出部分是灾后水体面积较灾前明显增大区域[①]

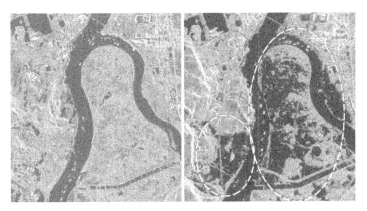

图 1-21 洪水发生前后雷达影像

无人机载 SAR 用于道路沉降和边坡变形监测,对预测边坡滑坡和崩塌风险具有非常重要的作用。道路和边坡含水量的差异会导致 SAR 微波反射率的不同,从而可以反演出这些区域的地质特性。再融合光学图像,通过标志点的检测实现边坡的三维测距,如图 1-22 所示。

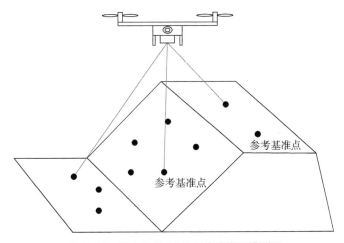

图 1-22 无人机载 SAR 融合光学三维测距

---

① 图来源:中国资源卫星应用中心,https://www.cresda.com/zgzywxyyzx/xxgk/kjjx/article/20240425140726594316658.html

通过以上几个例子的介绍，你们有没有觉得端侧模型优化是一件很有趣的工作？有点小期待吧！那么该如何开始这项有趣的工作呢？笔者认为编程是基础，打好编程基本功是第一步。比如能够精通 C 和 C++ 语言，理解汇编语言，会使用 Python 工具语言等都非常重要。我们要深刻明白：编程最终影响产品的质量！

举些例子说明编程基本功的重要性。

例子 1：堆、栈、BSS 段、数据段和代码段。

对于 C 程序进程的内存中堆、栈、BSS 段、数据段和代码段的分布（如图 1-23 所示），是一定需要深入去研究和理解的，不然在嵌入式开发中的表现就会很糟糕。

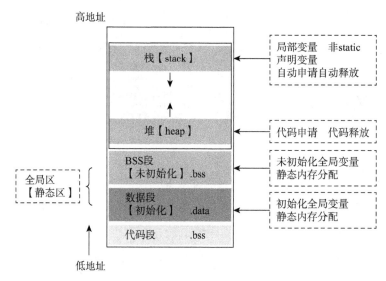

图 1-23　进程内存空间分布

我们一起看一下以下的基础代码来进一步讨论进程内存空间分配的问题（代码中有详细的注释）。

```
#include <stdlib.h>
//#include <memory.h>
//#include <stdio.h>

int a = 0;
// a 在全局区，加上初始化，所以分配在 .data 段

char *p;
// 指针 p 在全局区，因为未初始化，所以分配在 .bss 段

void main（）{
    int a0;
```

// 在函数体中定义的变量，编译后分配在栈区域

static int b = 0, c;
// static 声明并且初始化为 0 的 b，存储在 .data 段；static 声明但是没有初始化的 c，
// 存储在 .bss 段；b、c 均在全局区

char s [ ] = "Tensorflow";
// 定义一个字符数组，存放在栈区，字符串"Tensorflow"是在运行时赋值给数组 s
// 的。在栈中的数组比指针指向在其他区域内容的存取效率快

char *p1, *p2;
// 栈区域

char *p3 = "Pytorch";
// 等价 char *p3；p3 = "Pytorch"；
// 字符串"Pytorch"存放在常量区，p3 在栈上，p3 存放的是字符串"Pytorch"首
// 地址，"Pytorch"在编译时就赋值给 p3，p3 指向常量区
static int d = 0;
// 全局（静态）初始化区
p1 =(char*) malloc（5 * sizeof（char））;
p2 =(char*) malloc（10 * sizeof（char））;
// 分别分配 5 和 10 个字节的区域在堆区
// 头文件 #include "stdlib.h"
}

例子 2：内存拷贝，浅拷贝 / 深拷贝。

通过以上的讨论，相信读者朋友已经对内存分布有了深入的了解。我们再讨论另一个重要问题——内存拷贝，它分为浅拷贝 / 深拷贝。前者是指向同一块内存空间，而后者存放在不同的内存空间。这个在实际使用中需要格外注意，防止出现错误。

例子 3：快速排序。

如何把一串数据进行快速排序呢？在这里首先出场的就是冒泡排序，它是一个非常基础的排序算法，其在算法里的地位很像开始学习编程时的"Hello world"。我们用 C 语言实现的代码如下。

int array [ ] = {3, 0, 1, 4, 1, 5, 9, 2, 6};
int len = sizeof（array）/ sizeof（array [0]）;
// len 等于数组中元素个数，其中 sizeof（）是取字节操作符（关键字），不是函数，

// 它返回变量声明后所占的内存数,而不是实际长度;此外与之区别的 strlen() 是
// 函数,它求字符串从开始到遇到第一个 '\0' 的实际长度。
```
int* sort ( int array [ ], int len ) {
  int i;    // 比较轮数
  int j;    // 每轮比较的次数
  int tmp;  // 交换数据临时变量
  for ( i=0; i<len-1; ++i ) { // 比较 len-1 轮
    for ( j=0; j<len-1-i; ++j ) { // 每轮比较 len-1-i 次
      if ( array [j] < array [j+1] ) {
        tmp = array [j];
        array [j] = array [j+1];
        array [j+1] = tmp;
      }
    }
  }
  return array;
}
```

例子 4:类、继承、多态。

关于 C++ 代码中 const、volatile、restrict、static、大小端、类、基类、派生类、类的封装、继承(如表 1-2、表 1-3 所示)、多态、构造函数和析构函数[①]、抽象类、虚函数、纯虚函数、重载、友元函数等知识的理解和掌握很有必要。

表 1-2 访问权限

访问权限	public(公有)	protected(保护)	private(私有)
同一个类	Y	Y	Y
派生类	Y	Y	N
外部的类	Y	N	N
友元函数	Y	Y	Y

表 1-3 继承方式

继承方式	基类 public 成员	基类 protected 成员	基类 private 成员
public 继承	public 成员	protected 成员	不可见
protected 继承	protected 成员	protected 成员	不可见
private 继承	private 成员	private 成员	不可见

---

① 析构函数最好设计为虚函数

借助以下代码[①]和其中的注释，更便于理解。这段代码的编译命令为 g ++ test.cpp-o test。

```cpp
#include <iostream>
#include <stdlib.h>
#include <stdio.h>
#include <string.h>
using namespace std;

const double PI = 3.14159; // 只读变量
volatile int m; // m 是一个易变的位置，每次读取数据需要直接在内存中读取

// 类的三个基本特征：封装、继承、多态（覆盖，重载）
namespace dl
{
  namespace layer
  {
    /**
     * @brief Base class for layer. 基类
     *
     */
    // 类中包含纯虚函数的类为抽象类，抽象类不可以实例化对象
    class Layer
    {
    public:
        char *name; /*<! name of layer >*/

        static int size;
        // 静态成员（静态成员变量和静态成员函数）在类的所有对象中共享（创建
        // 多个类的对象，静态成员一个副本）
        // 如没有初始化，在创建第一个类的对象时，所有静态数据初始化为 0

        static int add（）
        {
```

[①] 这段代码是在 esp-dl/include/layer/dl_layer_base.hpp 的基础上进行的修改

```cpp
        cout <<" size" << size <<endl;
        return size;
    }

    /**
     * @brief Construct a new Layer object.
     *
     * @param name name of layer.
     */
    Layer(const char *name = NULL)
    {
        cout <<" Layer() " <<endl;
    }

    /**
     * @brief Destroy the Layer object. Return resource.
     *
     */
    ~Layer()
    {
    }

    // 纯虚函数
    virtual int bias()= 0;
protected:
    int width;
    int height;

private:
    int c;
};

// 派生类
class Conv2D : public Layer
{
```

```cpp
public:
    /**
     * @brief Construct a new Conv2D object.
     *
     */
    Conv2D()
    {
        cout <<" Conv2D() " <<endl;
    }

    /**
     * @brief Destroy the Conv2D object.
     *
     */
    ~Conv2D()
    {

    }

    // 重载+号运算符,属于成员函数,用于把两个Conv2D对象相加
    Conv2D operator+(const Conv2D& b)
    {
        Conv2D conv2d_;
    conv2d_.stride_y = this->stride_y + b.stride_y;
    // 类的对象有一个特殊的指针this,它指向对象本身
        conv2d_.stride_x = this->stride_x + b.stride_x;
        return conv2d_;
    }

    // 使用关键字friend修饰友元函数,重载-号普通函数,友元函数能访问类中
    // 的private或protected成员
    // 友元函数不在类里面,而是在类外,没有this指针
    friend Conv2D operator-(const Conv2D& a, const Conv2D& b)
    {
        Conv2D conv2d_;
```

```cpp
        conv2d_.stride_y = a.stride_y - b.stride_y;
        conv2d_.stride_x = a.stride_x - b.stride_x;
        return conv2d_;
    }

    int bias()
    {
        return 0;
    }
    //......

protected:
    int width;
    int height;
    //......

private:
    int stride_y;
    int stride_x;
    //......

};

} // namespace layer

} // namespace dl


using namespace dl;
using namespace layer;

int Layer::size = 6;
// static 成员变量如果类外初始化需要比类的对象先存在，所有类的对象共享一个
// 副本；
// 不和类的对象一样在堆栈中，而在静态存储区；
// 属于类，不属于对象，没有对象this指针

// union 共用体，成员共享同一块内存，大小由最大的成员大小决定，从低地址开始
```

// 存放
```cpp
typedef union {
  int i;
  float f;
} RangeData;

int main()
{
  RangeData d;
  d.i = 625;
  d.f = 3.14159;

  cout << sizeof(d) << endl;
  cout << d.i <<" " << d.f << endl;

  char * p1 =(char *) malloc(100*sizeof(char));
  char str1[100]={1, 2};
  const char str2[]="3.14159";
  cout <<" len1:" << strlen(p1) << endl;
  cout <<" len2:" << strlen(str1) << endl;
  cout <<" len3:" << strlen(str2) << endl;

  // int * restrict p2 =(int *) malloc(100*sizeof(int));
  // p 是访问 malloc 分配的内存唯一且初始的方式
  // Layer layer_; // 抽象类不可以实例化对象

  Conv2D conv2D_1;
  Conv2D conv2D_2;
  Conv2D conv2D_3;

  // 等价
  conv2D_3 = conv2D_1.operator+(conv2D_2);
  conv2D_3 = conv2D_1 + conv2D_2;

  conv2D_3 = conv2D_1 - conv2D_2;
  Conv2D::add();
  cout <<" main()" << endl;
```

```
    return 0;
}
```

鉴于本书篇幅限制，关于编程的重要性在这里只是列举几个简单的例子进行讨论。更多关于编程的知识，读者可以自行参考相关的书籍（关于代码风格，推荐大家阅读《Google C++ 编程风格指南》，将会有很多的收获），结合编程实践去深入理解。当然本书在后面的讨论中，也会重视编程的实践。再次强调一下笔者的观点：编程最终影响产品的质量！

1.3　端侧模型低功耗设计

端侧模型的功耗需要实现性能（Performance）和功耗（Power Consumption）的均衡。低功耗是我们的目标，它的好处有：

首先，针对用电池供电的端侧设备，可显著提高其续航时间。

其次，减少设备散热量，避免设备温度过高。

最后，优化后的产品稳定性更好，做到延长设备使用寿命和节约成本。

此外，由于散热变小，还可以让产品设计更加小型化。

下面讨论实现功耗优化的方法。

方法 1：算法优化

本书中讨论的模型压缩就是功耗优化的一方面。

方法 2：低功耗处理器

DSP 的功耗优于 GPU 和 CPU，在 DSP 满足需求的情况下，算法部署可以优先考虑 DSP。

对于多核异构处理器，采用异构计算（Heterogeneous Computing）实现多核的协同计算，做到算力、散热和能耗的均衡。

方法 3：动态切频实现处理器降频

处理器中 CMOS 电路的功耗和时钟频率呈线性关系：频率越低，其功耗越低。如图 1-24 所示，随着工作频率的增加，其电源电流也会增加（当然功耗也会增加）。在满足算法性能（包括时延）的前提下，通过动态切频实现处理器降频。

当算法不工作时，可以将处理器由工作状态变换为休眠状态，以更好地实现功耗优化。休眠状态包括浅度休眠和深度休眠，其中浅度休眠状态会关闭处理器核时钟，这样可以节省动态功耗[1]，而深度休眠状态不仅关闭处理器核时钟，还关闭电源，做到同时节省动态和静态功耗。

---

[1]　处理器功耗＝动态功耗＋静态功耗

Symbol	Parameter	Conditions	f_{HCLK}	Typ[1] All peripherals enabled[2]	Typ[1] All peripherals disabled	Unit
I_{DD}	Supply current in Run mode	External clock[3]	72 MHz	51	30.5	mA
			48 MHz	34.6	20.7	
			36 MHz	26.6	16.2	
			24 MHz	18.5	11.4	
			16 MHz	12.8	8.2	
			8 MHz	7.2	5	
			4 MHz	4.2	3.1	
			2 MHz	2.7	2.1	
			1 MHz	2	1.7	
			500 kHz	1.6	1.4	
			125 kHz	1.3	1.2	
		Running on high speed internal RC (HSI), AHB prescaler used to reduce the frequency	64 MHz	45	27	mA
			48 MHz	34	20.1	
			36 MHz	26	15.6	
			24 MHz	17.9	10.8	
			16 MHz	12.2	7.6	
			8 MHz	6.6	4.4	
			4 MHz	3.6	2.5	
			2 MHz	2.1	1.5	
			1 MHz	1.4	1.1	
			500 kHz	1	0.8	
			125 kHz	0.7	0.6	

1. Typical values are measures at T_A = 25 °C, V_{DD} = 3.3 V.

图 1-24　STM32 处理器在运行模式中电源电流值和时钟频率的关系（在 T_A=25 ℃、V_{DD}=3.3 V 条件下测量）[①]

1.4　端侧模型加固

模型部署在端侧，如何防止模型泄露是必须解决的重要需求。一般模型加固的方法有以下几种：

（1）把模型文件转换成 *.h 文件，集成到 C/C++ 工程里，再编译得到 *.so/*.lib 文件。其中 *.h 文件还需要添加混淆代码来增加破解难度。所谓代码混淆就是通过改写代码中的函数、类、变量为混淆符号，添加混淆逻辑等一系列措施，让代码难以理解。

（2）代码和模型混淆。

将函数名、变量名、参数名、类名以及模型中算子名称等做混淆处理，从而增加代码的逆向难度。读者朋友可以通过调研 OLLVM 代码混淆，做进一步研究，本书鉴于篇幅限制，不做进一步的讨论。

---

① 图中 Supply current in Run mode 译为运行模式中电源电流，All peripherals enabled 译为所有外设使能，All peripherals disabled 译为所有外设关闭，External clock 译为外部时钟，Running on high speed internal RC(HSI), AHB prescaler used to reduce the frequency 译为运行于高速内部时钟 RC 振荡电路产生的时钟信号，AHB 预分频器用于调整时钟频率。
图来源：STM32F103xC，STM32F103xD，STM32F103xE Datasheet-production data。

(3) AES 模型加密。

加密是重要的措施。加密分为对称加密和非对称加密，前者加密和解密采用同一个密钥，如 AES 加密；后者采用公钥和私钥两个密钥，如 RSA 加密。这两种加密流程如图 1-25 所示。

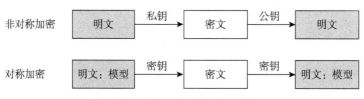

图 1-25 对称加密和非对称加密流程

AES 加密，它的密钥长度有三种，分别为 128 位（16 字节）、192 位（24 字节）、256 位（32 字节）。我们一般将模型文件采用 AES 算法加密，并且把密钥添加到代码中去编译。

在软件开发中，模型解密得到的模型文件一般需要使用诸如 SHA256 信息摘要算法进行完整性校验，避免模型解密不正确，导致下一步模型加载失败或者模型功能有误。

OpenSSL 库支持 AES、SHA256 等加密算法，支持软件工程开发。

(4) 加密锁，也叫加密狗（Dongle）。

加密锁的原理是软件在运行中需要和加密锁进行数据交互校验，如果没有交互成功，软件会采取必要的保护措施，如无法继续使用。加密锁是硬件设备，其防破解能力强。我们可以把密钥、模型的部分参数写到加密锁中，可以很好地保障模型的安全。

(5) 二进制应用加壳。

二进制应用加壳是在二进制应用程序的外面加入一段保护代码，当应用程序运行时优先获取程序控制权，保护程序不被逆向工程破解，从而保护核心算法等商业版权。

以上方法的应用，需要根据不同的产品去综合评估，以做到模型的加固。那么如何保证我们的模型加固达标呢？以下三个措施值得借鉴：

首先，代码审核机制让团队把好关。

其次，使用静态分析工具 IDA 等逆向分析工具评估分析。

最后，端侧系统安全测试、漏洞挖掘和分析等做好质量管控。

第 2 章

端侧 AI 芯片

"筚路蓝缕,以启山林"出自《左传·宣公十二年》,说的是先民驾柴车,穿着破烂衣服,披荆斩棘,开辟山林,可以想象出创业的艰辛。对于 AI 产品的研发,如 ChatGPT(用 ChatGPT 辅助写代码效率很高)、Vision Pro 头显、流媒体后视镜(具备宽动态、夜视能力)、内窥镜(包括高清影像 /4K、多模图像融合、蛇骨结构等)、机器人等等,再分解到模型核心算法研发、大模型推理部署加速优化、轻量化模型端侧部署、AI 集成优化,无不是汗水和智慧的结晶,研发人员不断付出再付出。对此更应明白"一饭一粥"来之不易,常怀感恩的心。

本章主要讲中央处理器体系结构、指令集架构、处理器架构、异构计算芯片,然后介绍端侧 AI 芯片(涉及芯片种类以及具体的芯片型号代表)。这些内容能够指导我们在端侧 AI 产品开发中该如何确定硬件方案(是采用 ARM,还是 DSP,或 GPU 等),以及如何进行算法的高性能优化。芯片是算力的载体,对人工智能的发展具有直接的推动力。

2.1 中央处理器体系结构

中央处理器(Central Processing Unit,CPU),是一块超大规模的集成电路,是计算机系统的运算和控制核心。按照中央处理器体系结构(Central Processor Architecture)分为:冯·诺依曼体系结构(Von Neumann Architecture)、哈佛结构(Harvard Structure)、并行处理结构、存算一体结构(Computing in Memory)。

冯·诺依曼体系结构(也叫普林斯顿结构,Princeton Architecture)将程序指令和数据存储空间统一编址(程序是特殊的数据,一样可以处理),也就是共用一套地址和数据总线。它设计简单,但是不能同时读取指令和操作数。

如图 2-1 所示,冯·诺依曼体系结构定义的计算机由运算器、控制器、存储器、输入设备和输出设备组成。其中运算器主要由算术逻辑单元(Arithmetic Logic Unit,

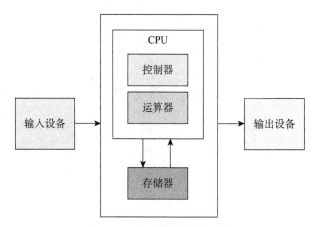

图 2-1 冯·诺依曼体系结构

ALU）、累加器、状态寄存器、通用寄存器等构成。运算器和控制器构成中央处理器，即 CPU。

Intel 的 X86 微处理器是冯·诺依曼体系结构的典型代表。

哈佛结构将程序指令和数据分开存储，可以同时读取指令和操作数，进而提高程序运行的访存效率。以 ARM（除 ARM7）和多数的 DSP 为代表。

AI 运算中数据频繁在内存和处理器之间搬运，计算效率和功耗是瓶颈。那么如何改进呢？

基于存算一体结构的芯片可以很好地解决这个问题。它将存储单元和处理单元设计在一个部件中，做到在数据存储过程中进行计算。相比较冯·诺依曼体系结构，它减少了数据在内存和处理器之间的搬运，降低带宽压力，也很好地解决了处理器处理速度与存储器数据读写速度数量级不对等的问题（俗称"内存墙"）。由于存储器数据读写的功耗远大于处理器处理功耗，所以这个结构也降低了整体的功耗。

2.2 指令集架构

在计算机系统中，用于指示处理器执行某种计算和控制的命令叫作指令。在该计算机系统中的全部指令的集合就叫作指令集，也叫指令集架构（Instruction Set Architecture，ISA）。指令集架构可分为 CISC、RISC、EPIC 和 VLIW 四类。

CISC（Complex Instruction Set Computer，复杂指令集计算机）的指令系统复杂庞大，长度不固定，指令格式多，寻址方式多，是早期的处理器指令集架构。其家族成员以 Intel、AMD 的 X86 为代表。

RISC（Reduced Instruction Set Computer，精简指令集计算机）大幅度简化指令集架构，仅保留常用的简单指令，通过组合简单指令实现复杂指令，其指令长度固定。其家族成员有 ARM、MIPS、RISC-V 等。

EPIC（Explicitly Parallel Instruction Computing，显示并行指令计算）从 VLIW 中衍生。采用 EPIC 指令集架构的处理器以 HP 和 Intel 合作研发的 Itanium（安腾）处理器为代表。[①]

VLIW（Very Long Instruction Word，超长指令字）是由美国 Multiflow 和 Cydrome 公司在 20 世纪 80 年代设计的指令集架构。VLIW 将多个相互独立和可以并行的指令打包成一个指令包（Packets），然后分配给超标量处理器来处理，从而实现并行执行多条指令。其中能够并行执行指令的条数等于 VLIW 指令集中的指令槽的个数。采用 VLIW 指令集架构的处理器以 Qualcomm 的 Hexagon DSP、ATI 的 GPU、TI 的 TMS320C62xx

---

① https://en.wikipedia.org/wiki/Explicitly_parallel_instruction_computing

为代表。如图 2-2 所示为 Hexagon V66[①] 处理器架构，其中指令排序器在每个指令周期[②]内并行执行包含 1～4 条指令的数据包。图中 S0、S1、S2、S3 分别代表 VLIW 指令集的 4 个指令槽对应的执行单元，图中四个虚线框中的不同类型指令说明不同的指令槽中特定的类型指令才能组合为一个包并行执行。指令的打包是由编译器实现的，所以编译器的优化能力决定指令的吞吐量。

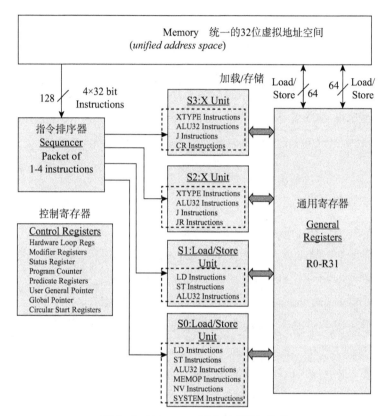

图 2-2　Hexagon V66 处理器架构[③]

2.3　处理器架构及异构计算芯片

先有以上所述的指令集架构，然后才有处理器架构，就好比前者是设计图纸，后者是按照图纸建造的大楼。处理器架构是处理器的设计结构和组织方式，决定了硬件如何

---

① 其中 V66 是 DSP 架构版本。
② 指令周期（Instruction Cycle）是指处理器完成取指令（Instruction Fetch，IF）、指令译码（Instruction Decode，ID）、执行指令（Execute，EX）的全部时间。时钟周期（cycle，clock cycle）为处理器主频的倒数，是最最小的时间单位。
③ 图来源：Qualcomm Hexagon V66 Programmer's Reference Manual. 80-N2040-42 A. November 17，2017

执行软件指令以及如何处理和存储数据。指令集架构的先进性直接影响处理器的性能。目前主流的指令集架构以 X86、ARM、MIPS、RISC-V 为代表，它们相对应的处理器架构具体讨论如下。

X86 指令集源于 Intel 开发的处理器所采用的一种复杂指令集且闭源，其特点有通用性强、兼容性好、功能强大。后来的 AMD 等公司处理器均支持 X86 架构。X86 指令集架构在服务器领域中占主导地位。

ARM 是 32 位精简指令集架构，其指令集闭源，以授权为主，授权方式包括指令集架构授权、IP 核授权等。由于其低功耗、低成本特性，非常适合嵌入式领域，目前在移动处理器市场中占主流，比如华为、Google、Qualcomm、Apple、MTK 的处理器架构均采用其授权的指令集架构。

MIPS 指令集，源于 20 世纪 80 年代初斯坦福大学 Hennessy 教授的研究团队。它是一种精简指令集，具有固定的指令长度（32 位），且指令格式简单。值得关注的是：MIPS 所属公司宣布投入 RISC-V 阵营。[1]

RISC-V（发音为"risk-five"），源于 2010 年加州大学伯克利分校的一个项目，其指令集完全开源、简化流程、易于移植、模块化设计。在 IoT、AI 场景中具有非常大的发展潜力。[2] 阿里平头哥玄铁 910、C906、E906 处理器采用 64bit RISC-V 架构，分别应用到高性能、高能效、低功耗场景，此外全志 D1s、D1-H 芯片是基于平头哥 C906 IP 核，广泛应用在 AIoT 场景。

异构计算芯片是芯片架构的一种趋势。所谓异构计算芯片就是把 CPU、NPU、DSP、GPU、AI 加速器、协处理器、FPGA 等不同类型的微处理器和微控制器的 IP 内核通过优化设计和 3D 封装工艺集成为一个芯片，实现通用计算和高性能、低功耗计算。如全志 V853 芯片（全志 V853 是一款高性能、低功耗的新一代边缘 AI 视觉处理芯片，可广泛用于安防摄像头、智能后视镜、扫描翻译笔、智能门锁、智能考勤门禁、网络摄像头、运动相机、行车记录仪等智能化场景）集成 ARM Cortex-A7、RISC-V E907 和 NPU，系统架构如图 2-3 所示：

---

[1] We're taking RISC-V to new heights. With a long history of developing RISC architectures for industrial-strength products, today MIPS is accelerating RISC-V innovation for a new era of heterogeneous processing. 译：我们正在将 RISC-V 推向新的高度。凭借为工业级产品开发 RISC 架构的悠久历史，如今 MIPS 正在加速 RISC-V 的创新，进入异构处理的新时代。摘自：https://www.mips.com/。

[2] https://riscv.org/about/

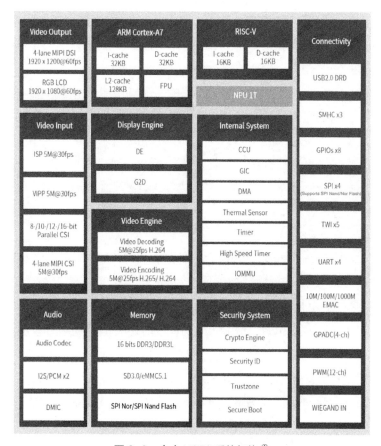

图 2-3　全志 V853 系统架构[1]

XBurst 系列 T41 芯片集成 XBurst2 双核 CPU、RISC-V 协处理器、AI Engine、AIP 等核，如图 2-4 所示。

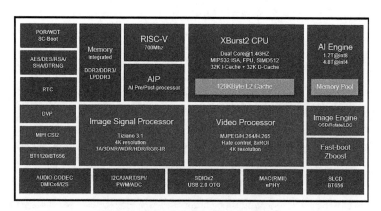

图 2-4　T41：普惠 AI 视频处理器[2]

---

[1]　图来源：https://v853.docs.aw-ol.com/
[2]　图来源：http://www.ingenic.com.cn/products-detail/id-2.html

2.4 端侧 AI 芯片

端侧 AI 芯片是实现端侧设备 AI 计算和处理的芯片。这是一个宽泛的概念，可以把能够满足当前 AI 算法需求的芯片叫作端侧 AI 芯片，其目标是追求体积、功耗、性能和成本的平衡，主要有 ARM、DSP、GPU、FPGA、ASIC、NPU，还有前沿研究芯片——类脑芯片。

下面列出汽车行业主要公司的汽车 AI 芯片代表（后期新的公司和新推出的芯片，读者朋友可以在这个基础上关注和研究）：

Qualcomm：602A、820A、QCM6125、QCM6490、SA6350P、SA6155P、SA8155P、SA8195P、SA8295P（汽车座舱 SoC 芯片，其中 SA8155P 对应移动手机平台 Snapdragon 855）、Snapdragon Ride Flex（用于数字座舱、辅助驾驶、自动驾驶等）等；

NVIDIA：Tegra KI/XI/Parker、Xavier、Orin、AGX Pegasus、Atlan 等；

Mobileye：EyeQ3/Q4/Q5/Q6 等；

AMD：Ryzen V1000；

地平线（Horizon Robotics）：征程 2/3/5 等；

华为：昇腾 310/910（HUAWEI Ascend 910 基于华为自研达芬奇架构 3D Cube 技术，性能为 320 TFLOPS @FP16，640 TOPS[①] @INT8）、麒麟 710A/990A 等；

Tesla：HW3.0/4.0、DOJO 等。

2.4.1 ARM

ARM 架构是目前在移动端设备和嵌入式系统中占据主导地位的处理器架构，它包括指令集架构和微架构（Micro-Architecture）两个方面。ARM 指令集架构定义了指令集、寄存器组、异常模型、内存模型，以及调试、跟踪和分析等[②]。

ARM 架构按照版本可分为 Armv7（ARM 公司在 2006 年推出）、Armv8（2011 年推出）和 Armv9（更强的安全性，引入新的向量处理扩展 SVE2，向后兼容 Armv8）。其中 Armv8 包括 A（Armv8-A）、R（Real-Time）和 M（Microcontroller）系列。Armv8-A 架构包括 32 位和 64 位执行状态，每个状态都有自己的指令集，分别为 Arch32 和 AArch64。Arch32 描述 Armv8-A 架构的 32 位执行状态，几乎与 Armv7 架构相同。AArch64 用于描述 Armv8-A 架构的 64 位执行状态。

微架构是指令集架构的硬件实现。微架构的内容包括流水线长度和布局、缓存的数量和大小、单个指令的周期计数和可实现的可选功能等。举例说明一下：Cortex-A53 和

---

① TOPS：万亿次运算

② 参考资料 https://developer.arm.com/documentation/102404/0201/?lang=en

Cortex-A72 均是基于 Armv8-A 架构的实现（相同 ARM 架构），但它们的微架构不同，详见图 2-5。

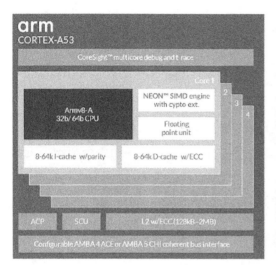 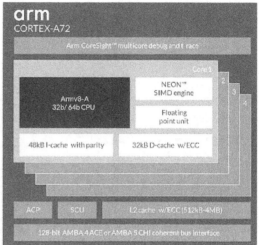

图 2-5　同 ARM 架构，不同微架构[1]

ARM 处理器芯片[2] 根据其中的 CPU 型号可分为 ARM7、ARM9、ARM9E、ARM10E，再到 ARM11 处理器改用 Cortex 命名，分为 A、R 和 M 三个系列[3]，其中具体如下：

Cortex-A 面向尖端基于虚拟内存的操作系统和用户应用；

Cortex-R 针对实时系统；

Cortex-M 为微控制器。

列举几个代表说明一下：

STM32 系列 32-bit 微控制器是基于 Cortex-M 的微控制器，如图 2-6 所示为其各个系列芯片。

---

[1] 图来源：https://developer.arm.com/documentation/102404/0201/Architecture-and-micro-architecture?lang=en

[2] 更多资料参考 https://www.arm.com/architecture/cpu

[3] 参考资料 https://developer.arm.com/ip-products/processors/cortex-a，https://developer.arm.com/ip-products/processors/cortex-r

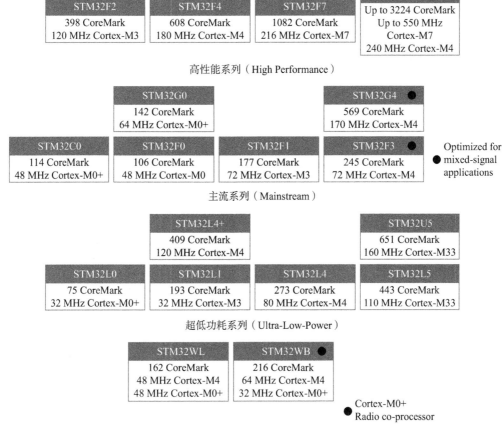

图 2-6 STM32 系列微控制器[1]

RK3588 芯片搭载 4×Cortex-A76+4×Cortex-A55 8 核 CPU 和 ARM Mali-G610 MC4 GPU（专用于 2D 图形加速模块），内置 6TOPS 算力 NPU。在智能座舱、车载仪表盘等领域，凭借其超高运算性能，可实现一芯多屏。用一颗 RK3588M 同时驱动车载信息娱乐系统、液晶仪表板、电子后视镜、后排头枕屏等多块屏幕，同时支持 360° 环视功能，形成可靠安全的智能网系统，给用户带来高科技感交互体验。[2]

RK3399 芯片中 CPU 部分为双 Cortex-A72+4×Cortex-A53 大小核，频率最高达 1.8GHz 的低功耗、高性能应用处理器。

RV1126 芯片集成四核 ARM Cortex-A7 和 RISC-V MCU，内置 2T 算力 NPU。它的系统架构如图 2-7 所示。该芯片广泛应用在车载摄像头（ADAS、DMS、BSD 等功能）、车载电子、行车记录仪、智能门锁、智能门铃、网络摄像头（IPC）、安防监控、机器人等场景。

---

[1] 图来源：https://www.st.com/content/st_com/zh.html
[2] 参考资料：https://www.rock-chips.com/a/cn/news/rockchip/2022/1209/1750.html

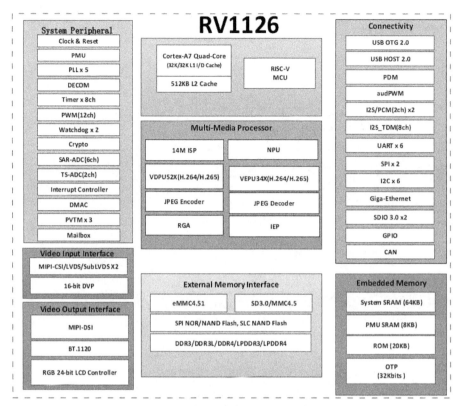

图 2-7 RV1126 系统架构[1]

NXP 的 i.MX 系列产品包括 i.MX RT 系列、i.MX 9 系列、i.MX 8 系列、i.MX 7 系列、i.MX 6 系列、i.MX28 系列[2]。其中 i.MX 6ULL 处理器搭载单个 Cortex-A7 内核，运行速度高达 900 MHz，架构如图 2-8（a）所示。它被广泛应用在物联网网关、终端节点和消费类电子产品等场景。i.MX 6Quad 处理器搭载 4 个 Cortex-A9 内核，每个内核运行频率高达 1.2 GHz，其架构如图 2-8（b）所示。它被应用在多摄像头环视、感测泊车辅助系统、中端到高端车载信息娱乐系统、电子驾驶舱等场景。

---

[1] 图来源：Rockchip RV1126 Datasheet
[2] 更多资料可以参考：https://www.nxp.com.cn/products/processors-and-microcontrollers/arm-processors/i-mx-applications-processors:IMX_HOME

第 2 章 端侧 AI 芯片

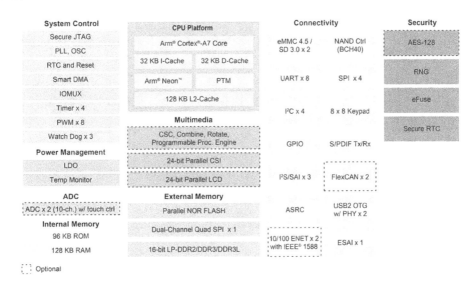

（a）i.MX6ULL 处理器架构

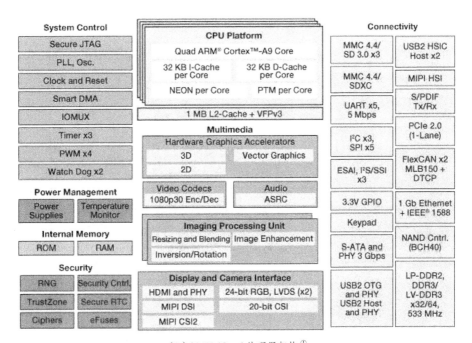

（b）i.MX 6Quad 处理器架构[①]

图 2-8　i.MX 系列系统架构

---

① 图来源：https://www.nxp.com.cn/products/processors-and-microcontrollers/arm-processors/i-mx-applications-processors/i-mx-6-processors/i-mx-6ull-single-core-processor-with-arm-cortex-a7-core:i.MX6ULL，https://www.nxp.com.cn/products/processors-and-microcontrollers/arm-processors/i-mx-applications-processors/i-mx-6-processors/i-mx-6quad-processors-high-performance-3d-graphics-hd-video-arm-cortex-a9-core:i.MX6Q

2.4.2 DSP

DSP（Digital Signal Processor，数字信号处理器）是一种专门用于实时数字信号处理的微处理器，适合仪器仪表、工业控制、音视频处理[①]、通信、雷达、医疗设备（如心电图监测、CT、超声检测设备、助听器等）、图像处理和计算机视觉[②]等领域。DSP 的主要特点一般有以下几点：

（1）高性能

采用专用硬件乘法器实现乘法操作和乘累加，而且在单个指令周期内完成一次乘法和一次加法；

DSP 设计有很多特殊针对数字信号处理运算的指令，如图 2-9 所示的复数乘法指令用于支持 FFT 计算。

[①] 音频算法相关模块介绍

音频算法是研究和实现声音信号处理的一门技术。目前相关的软件框架有 Kaldi（用 C++ 开发的语音识别工具）、WebRTC、Speex、Audacity 等。它的研究方向有很多，以下主要介绍语音增强、语音交互、音频编解码和声纹识别等。

a. 语音增强

语音增强（Speech Enhancement）实现将带有噪声的声音中的有用信号加强，它包括以下内容：

a）音频 3A 算法包括回声消除（AEC）、噪声抑制（ANS）和自动增益控制（AGC），其中噪声抑制包括主动降噪和啸叫抑制（AFC），这些算法可以通过调用 SpeexDSP 库或 WebRTC 中的相关模块来测试评估。

b）音频滤波器实现对音频信号的频率进行低通/高通/带通/带阻等处理，涉及的技术有 FIR（Finite Impulse Response，有限长单位冲激响应）/IIR（Infinite Impulse Response，无限脉冲响应）滤波器、均衡器、自适应滤波器以及频谱分析等。

c）音效处理包括虚拟环绕声、串扰消除、头中效应消除、混响去除等。

d）麦克风阵列音频处理包括波束形成、盲源分离、声源定位等。

b. 语音交互

a）ASR（Automatic Speech Recognition，自动语音识别）实现将语音转换成文本，涉及的技术有语音检测（VAD）、唤醒词检测（KWS）、声学模型和语言模型等。

b）TTS（Text To Speech，从文本到语音）实现将文本转换成语音，让机器说话，涉及的技术有声码器（用于合成语音）、个性化/情感化 TTS 等。

c）NLP（Natural Language Processing，自然语言处理）实现计算机对文本的理解。例如出色的聊天机器人 ChatGPT 是 NLP 的一大突破。

c. 音频编解码

EVS（音频编码器）、G.7xx、AAC（高级音频编码，有损压缩）、Opus 等。

d. 声纹识别（Voiceprint Recognition）

声纹识别属于生物识别技术的范畴，它通过对声音的特征参数提取和匹配实现声音个体的区分。

[②] 人脸识别、OCR、图像分类、目标检测、语义分割以及底层视觉（Low-Level Vision，如视频去抖动、超分、去模糊、视频插帧、拍照夜增强、暗光增强、去雾、去噪等）。

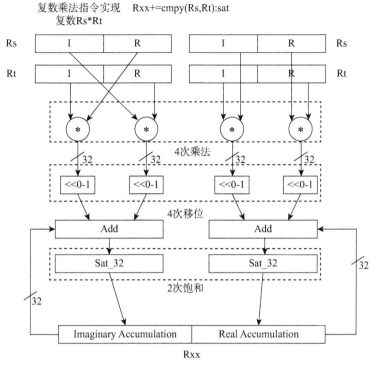

图 2-9 Hexagon 中复数乘法指令 [1]

采用哈佛总线结构,其中数据和程序具有各自的存储空间,数据总线和程序总线分开工作,可以同时进行寻址,进而提高数据吞吐量和程序处理能力;

指令周期的三个阶段:取指、译码、执行,采用流水线操作;

执行硬件多线程。

(2)低功耗

芯片设计中做到性能和功耗的均衡优化。一般 DSP 核运行在低的时钟频率,采用 SIMD 扩展指令集,实现更低的功耗,特别适合无线通信、移动设备中电池的续航,以及设备发热、卡顿问题的解决。

(3)可编程灵活性

首先,支持高性能乘法和累加运算,支持并行运算等特性,能够满足不同的算法需求;其次,对比 FPGA 等芯片,更加容易编程。

(4)低成本

目前市场上有很多芯片厂商提供 DSP 芯片,下文主要介绍一些主流的 DSP 芯片,其中对 Hexagon 做重点介绍,便于读者朋友对 DSP 的硬件架构有更深的认识。

首先是德州仪器(Texas Instruments,TI)推出的四大系列 DSP:

---

[1] 图来源:Qualcomm Hexagon V66 Programmer's Reference Manual. 80-N2040-42 A. November 17, 2017

a. TMS320C2000 系列，代表芯片有 F28377、F280049、F28335、F28035，其特点是定点 DSP，用于数字控制系统；

b. TMS320C5000 系列，分为系列 C54xx、C55xx，代表芯片有 C5402、C5505，其特点是定点 DSP，适合低功耗通信终端；

c. TMS320C6000 系列，其中定点系列是 C62xx、C64xx，浮点系列是 C67xx，代表芯片有 C6416（定点）、C6713（浮点）、C6748（浮点），其特点是高性能，适合高性能多功能复杂应用系统，如机器视觉、视频监控、医学成像等；

d. OMAP 系列（Open Multimedia Application Platform，开放式多媒体应用平台），集成 ARM 处理器和浮点 DSP 核，代表芯片有 OMAP-L132、OMAP-L138，多应用在移动多媒体信息处理及无线通信等设备。

ADI 公司的 DSP 系列主要有以下几种：

a. ADSP-21xx 系列定点 DSP，代表芯片有 ADSP-21990，适合语音处理等场景；

b. Blackfin 嵌入式处理器，代表芯片有 ADSP-BF701、ADSP-BF533、ADSP-BF532、ADSP-BF531，适合音频、视频、图像处理；

c. SigmaDSP，是可编程的单芯片音频 DSP，代表芯片有 AD1940、AD1941、ADAU1701、ADAU1702、ADAU1787、ADAU1401 等，适用汽车和便携式音频产品。[1] 如图 2-10 所示是采用 ADAU1787 芯片方案实现的主动降噪（ANC）耳机，高效降噪和低功耗性能给音乐和电影爱好者带来优质的声音体验。

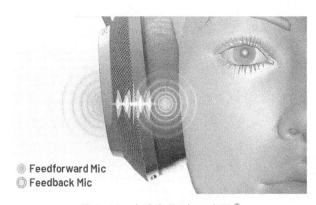

图 2-10　主动降噪耳机示意图 [2]

AT&T 公司开发的 DSP16/32 系列 DSP。

NXP（2015 年 NXP 收购由 Motorola 创立的 Freescale 半导体）的 56F81xxx 系列、56F82xxx 系列、56F83xxx 系列、56F84xxx 系列的 DSP，如图 2-11 所示。

---

[1] 参考资料 https://www.analog.com/cn/product-category/processors-dsp.html

[2] 图来源：https://www.analog.com/cn/signals/articles/bowers-wilkins-noise-canceling-headphones.html

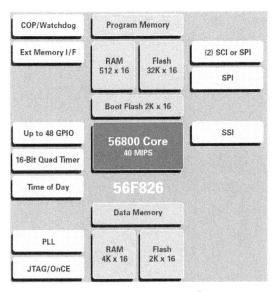

图 2-11　DSP56F826 [1]

Cadence Tensilica HiFi DSP 是音频信号处理 DSP，可用于满足音频和语音信号处理等需求，例如编解码、语音活动检测（VAD）、关键词识别（KWS）、AI 降噪等。它具有高性能和低功耗的特性，广泛应用在 TWS 蓝牙耳机、助听器、智能手环、汽车信息娱乐等。其中 HiFi 3z DSP 的系统架构如图 2-12 所示。

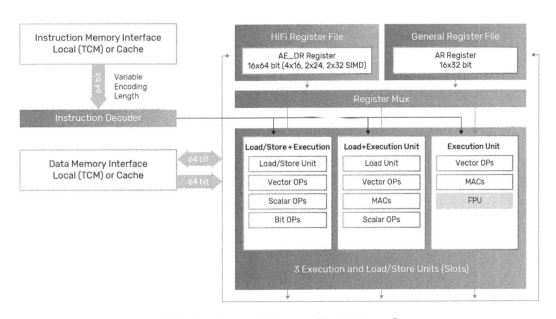

图 2-12　Cadence® Tensilica® HiFi 3z DSP [2]

---

[1]　图来源：https://www.nxp.com.cn
[2]　图来源：https://www.cadence.com/zh_TW/home/tools/ip/tensilica-ip/hifi-dsps/hifi-3.html

Cadence Tensilica HiFi DSP 是一款音频信号处理 DSP，可用于满足音频和语音编解码等需求。它具有小尺寸和低功耗的特性，广泛应用在 TWS 蓝牙耳机、助听器、智能手环等小型电子设备中。

全志 R329 芯片集成双核 HiFi4@400MHz，如图 2-13 所示，广泛应用在智能语音市场。

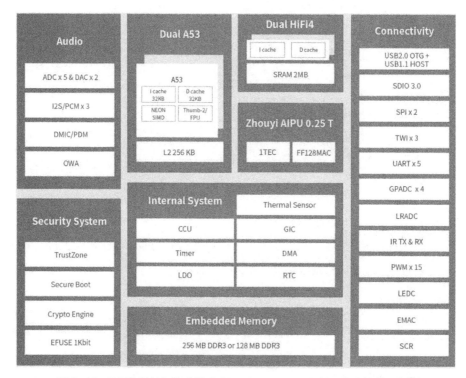

图 2-13　R329 芯片架构（集成 CPU、DSP、NPU 的异构多核芯片）①

Kalimba DSP 是 Qualcomm 的单核 120MHz 音频 DSP，集成在 QCC514x、QCC304x、CSR 系列（如 CSR 8675）等芯片中，它实现高性能和高品质音频功能，广泛应用在 TWS 蓝牙耳机等场景。

Hexagon 是 Qualcomm 的"第六代数字信号处理器"，也称为 QDSP6，是骁龙处理器（Snapdragon）中的 DSP 核。如图 2-14 所示为骁龙处理器 SM8150 的芯片框图，它有四个 DSP 核，具有各自的应用功能，分别为传感器（sDSP）、调制解调器（mDSP）、音频（aDSP）、计算（cDSP），一般 cDSP 就是 Hexagon。

---

① 图来源：https://www.aw-ol.com/chips/4

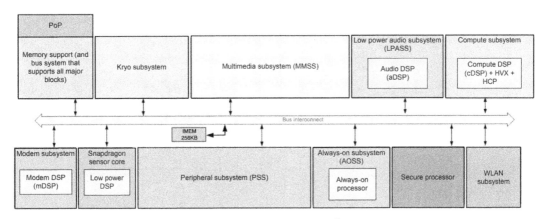

图 2-14　SM8150 框图[1]

下表是目前 Snapdragon 主流的处理器型号。其中 Hexagon Tensor Accelerator 为 Tensor 加速器，Hexagon Vector eXtensions 为矢量扩展，Hexagon Scalar Accelerator 为 Scalar 加速器。

表 2-1　Snapdragon 主流处理器型号[2]

	SM8350	SM8450	SM8475	SM8550
	Snapdragon 888	Snapdragon 8 Gen 1	Snapdragon 8+ Gen 1（8 Gen 1 Plus）	Snapdragon 8 Gen 2
CPU	Kryo 680 CPU	Kryo CPU	Kryo CPU	Kryo CPU
GPU	Adreno 660 GPU	Adreno 730 GPU	Adreno 730 GPU	Adreno 740 GPU
Hexagon（DSP）	• Hexagon Tensor Accelerator • Hexagon Vector eXtensions（HVX） • Hexagon Scalar Accelerator	• Hexagon Tensor Accelerator • Hexagon Vector eXtensions • Hexagon Scalar Accelerator • 支持混合精度：INT8+INT16 • 支持单精度：INT8、INT16、FP16	• Hexagon Tensor Accelerator • Hexagon Vector eXtensions • Hexagon Scalar Accelerator • 支持混合精度：INT8+INT16 • 支持单精度：INT8、INT16、FP16	• Hexagon Tensor Accelerator • Hexagon Vector eXtensions • Hexagon Scalar Accelerator • 支持混合精度：INT8+INT16 • 支持单精度：INT4、INT8、INT16、FP16

---

[1] 图来源：Qualcomm/Hexagon_SDK/3.5.4/docs/images/80-VB419-108_Hexagon_DSP_User_Guide.pdf
[2] 参考资料：https://www.qualcomm.com/products/mobile/snapdragon/smartphones/snapdragon-8-series-mobile-platforms

Hexagon 分为主处理器（Hexagon core）和可选的 SIMD 协处理器（SIMD coprocessor）两个部分，前者主要负责标量计算，后者协处理器指 HVX（Hexagon Vector eXtensions，Hexagon 矢量扩展内核），主要负责矢量计算。

主处理器包含 4 个或更多的硬件线程，每个硬件线程均有 32 个 32bit 通用寄存器（R0—R31）、4 个 32bit 预测判断寄存器（P0—P3）[1]、4 个执行单元（S0—S3），如图 2-15 所示。

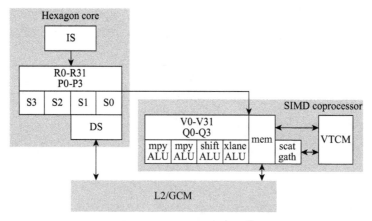

图 2-15　Hexagon DSP[2]

协处理器 HVX 包含 32 个 1024bit（128B）的矢量寄存器（V0—V31），4 个 128bit 预测寄存器（Q0—Q3）、4 个执行单元（ALU）、内存读写单元（mem）、scatter-gather（非连续内存操作）。HVX 可实现 1024bits 数据的并行处理，实现高性能、低功耗。例如指令：V1 = vmem（R0）实现从通用寄存器 R0 指向的地址空间（Memory addresses）加载 128 个字节到矢量寄存器 V0 中。

协处理器 HVX 有 4 个 HVX context，并且共享 L2 Cache 和 VTCM 内存。图 2-16 所示为 Hexagon 的两级缓存机制，其中 L1 仅能被标量单元访问，L2 是标量单元的二级缓存，是协处理器 HVX 的一级缓存。

---

[1] 属于控制寄存器（Control registers）
[2] 图来源：Qualcomm Hexagon V66 HVX Programmer's Reference Manual 80-N2040-44 Rev. B October 15, 2018

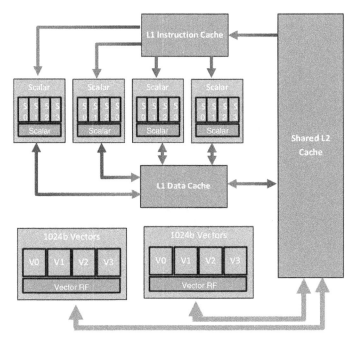

(a) Hexagon 内存子系统

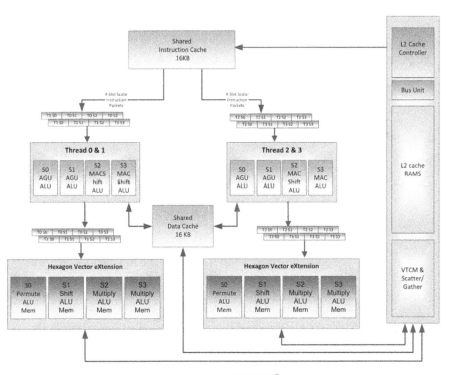

(b) Hexagon V66 框图 [1]

图 2-16　Hexagon 的两级缓存机制

---

① 图来源：Qualcomm/Hexagon_SDK/3.5.4/docs/images/80-VB419-108_Hexagon_DSP_User_Guide.pdf

2.4.3 GPU

GPU（Graphics Processing Unit，图形处理器），GPU 拥有大量的算术运算单元（ALU）和超长的流水线，适合并行计算。芯片有 Mali GPU（ARM® Mali™ GPU 系列[①]）、PowerVR GPU、Adreno GPU（高通 Adreno 系列 GPU）和 Nvidia GPU。以下列举几个芯片作为代表进行说明。

RK3399 芯片，其 GPU 核型号是 Mali-T860 GPU，支持 OpenGL ES1.1/2.0/3.0/3.1、OpenCL1.2。

RK3588 芯片，其 GPU 核型号是 Mali-G610 MC4，支持 OpenGL ES1.1/2.0/3.1/3.2、Vulkan 1.1/1.2、OpenCL 1.1/1.2/2.0，内嵌高性能 2D 图像加速模块。

Jetson 平台是 Nvidia AI 边缘平台，它包括 Jetson Orin、Jetson Xavier、Jetson TX2、Jetson Nano 等系列。

2.4.4 FPGA

FPGA（Field Programmable Gate Array，现场可编程门阵列）是在 PAL（Programming Array Logic，可编程阵列逻辑）、GAL（Generic Array Logic，通用阵列逻辑）、CPLD（Complex Programming Logic Device，复杂可编程逻辑器件）等可编程器件基础上发展的新型数字集成电路芯片。FPGA 是专用集成电路（Application Specific Integrated Circuit，ASIC）领域中一种半定制电路，其通过硬件编程（区别于软件编程）改变芯片内部电路结构，从而满足不同硬件的要求。FPGA 的优势如图 2-17 所示：

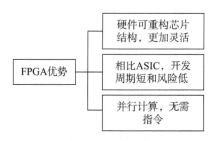

图 2-17　FPGA 优势

FPGA 的编程语言也叫硬件描述语言（Hardware Description Language，HDL），有 Verilog HDL、VHDL（Very-High-Speed Integrated Circuit Hardware Description Language，超高速集成电路硬件描述语言）、SystemVerilog（SV 语言），其中对新手一般推荐学习 Verilog HDL。目前市场上供应 FPGA 的公司有 Xilinx（赛灵思，目前已经被 AMD

[①] 更多资料可以参考：https://developer.arm.com/documentation/#&cf［navigationhierarchiesproducts］=%20IP%20Products，Graphics%20and%20Multimedia%20Processors，Mali%20GPUs

收购)、Altera(阿尔特拉,已被Intel收购)、Microsemi(美高森美)、Lattice(莱迪思)等。

proFPGA是一个新推出的具备可扩展、模块化的多FPGA方案。Xilinx的ProFPGA XCVU440 FPGA提供多达30 M个ASIC门(一个FPGA中),10个达1327个用户I/O的扩展站点和48个高速串行收发器(MGT)。[①]

Xilinx的Zynq™ 7000 SoC系列芯片采用ARM+FPGA架构,该系列有单核Zynq7000S和双核Zynq-7000。[②]其中Zynq-7000芯片集成双核ARM Cortex-A9处理器和Artix7或Kintex™ 7可编程逻辑,架构如图2-18所示,可应用在多摄像头驾驶员辅助系统、医疗内窥镜、机器视觉、4K2K超高清电视等嵌入式应用场景。

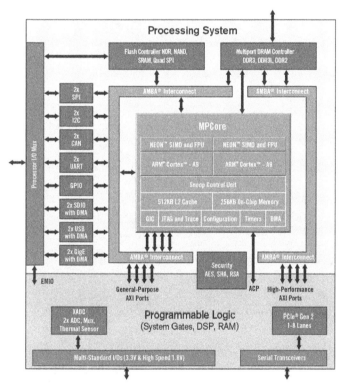

图2-18 Zynq-7000芯片架构[③]

2.4.5 NPU

NPU(Neural Network Processing Unit,神经网络处理器)属于专用定制芯片ASIC

---

① 参考 https://china.xilinx.com/products/boards-and-kits/1-66ql3z.html, https://www.profpga.com/products/fpga-modules-overview/virtex-ultrascale-based/profpga-xcvu440
② 参考 https://china.xilinx.com/products/silicon-devices/soc/zynq-7000.html#productAdvantages
③ 图来源:包含软硬件可编程功能的SoC, https://www.amd.com/zh-cn.html

的一种。NPU 是专用于人工智能硬件加速的微处理器，其在性能、功耗、面积方面具有很大的优势。NPU 芯片有华为 DaVince 架构 NPU（昇腾 AI 芯片）、谷歌 TPU（Tensor Processing Unit，张量处理器）、阿里达摩院含光 NPU（AliNPU）、寒武纪 NPU 等。

V853 芯片中集成的 NPU 核（Vivante 神经网络处理器 IP），其系统架构如图 2-19 所示，包括可编程引擎（Programmable Engines，PPU）、神经网络引擎（Neural Network Engine，NN）和高速缓存。其中神经网络引擎包含 NN 模块和 Fabric（Tensor Process Fabric，TPF）。NN 模块用于卷积计算操作，Fabric 用于高速数据交换。NPU 处理性能最大达到 1T 算力。NPU 支持 OpenCL、OpenVX、Android NN、ONNX 等 API，支持 PyTorch、TensorFlow、TFLite、ONNX 等深度学习模型的加载，支持 UINT8、INT8、INT16 数据类型。

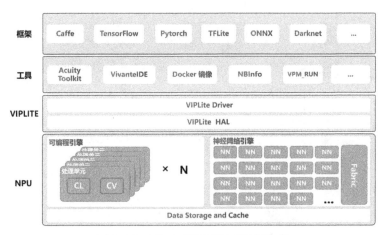

图 2-19　V853 芯片 NPU 系统架构[1]

RK3588 中 NPU 核处理性能达到 6TOPS 算力，支持 INT4、INT8、INT16、FP16、BF16、TF32 数据类型。

RV1126 内置 2T 算力 NPU，支持 INT8、INT16 混合精度，其中内置 ISP 模块支持最大像素为 14M，像素尺寸为 ≤4416×3312。如图 2-20 所示为其 NPU 架构，包括 HIF（Host Interface）[2]、PM（Power Management）、NN Engine（Neural Network Engine）[3]、VPU（Vector Processing Unit）[4]。

---

[1] 图来源：https://v853.docs.aw-ol.com/npu/dev_npu/
[2] AHB（Advanced High-performance Bus，高级高性能总线），AXI（Advanced eXtensible Interface，高级可拓展接口）
[3] NN 引擎是神经网络算法的主要处理单元，实现并行卷积计算，以及多种激活函数，如 leaky_relu、relu、relu1、relu6、sigmod、tanh 等。
[4] 矢量处理单元为可编程 SIMD 处理器单元，支持 OpenCL。VPU 提供高级图像处理功能，例如在一个周期中，VPU 执行一条 MUL/ADD 指令或两个 16 分量值的点积。大多数元素运算和矩阵运算在 VPU 中处理。

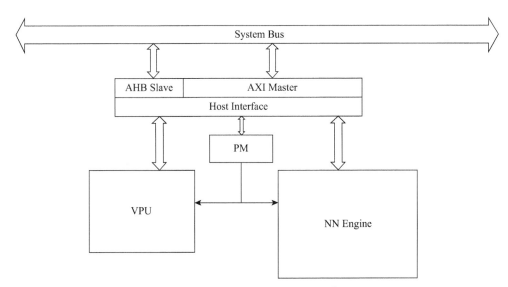

图 2-20　RV1126 NPU 架构[①]

2.4.6　类脑芯片

类脑芯片，顾名思义就是模拟人脑神经元机制的芯片。类脑芯片的优势在于计算能耗、能力、效率等方面有大幅度提升。类脑芯片的代表有 IBM TrueNorth、Intel Loihi、Qualcomm Zeroth 等，它们均是基于脉冲神经网络。其中 TrueNorth 芯片把存储单元作为突触，计算单元作为神经元（100 万个），传输单元作为轴突（25600 万个），总计集成 4096 个并行和分布式神经内核，用了 54 亿个晶体管，功耗却只有 70mW，像一台邮票大小的神经突触超级计算机，可以使用智能手机的电池运行一周。[②]

---

① 图来源：Rockchip RV1109/RV1126 Technical Reference Manual
② 参考 https://www.ibm.com/blogs/research/2016/12/the-brains-architecture-efficiency-on-a-chip/，https://www.intel.cn/content/www/cn/zh/research/neuromorphic-computing.html

第 3 章

Linux 开发环境及工具介绍

本章主要介绍 Shell 命令解释器和脚本、环境变量配置、编译器 GCC、CMake 项目构建、Make 编译、LLVM 和 Clang、Android SDK 和 NDK 配置、Clang/Clang++ 编译 C/C++/Neon 程序和终端运行、GDB 调试及 ADB 工具等。"工欲善其事，必先利其器"正是本章内容的意义所在。

3.1 Shell 命令解释器和脚本

Shell（壳）包括命令解释器和脚本。其中 Shell 命令解释器用于解释用户输入的命令，提供用户与系统内核交互的接口。Shell 命令解释器有 /bin/sh（Bourne Shell）、/bin/bash（Bourne Shell 的替代品）等。Shell 脚本（Shell Script）是由一连串的 Shell 命令组成的脚本语言，它通过 Shell 命令解释器解释执行，而不需要编译。Shell 脚本文件后缀名为 .sh，Shell 脚本文件第一行一般为 #!/bin/bash，用来指定 Shell 命令解释器为 /bin/bash。Shell 脚本一般应用在重复性操作、批量操作、交互任务等中。Shell 架构如图 3-1 所示。对初学者来说，建议先入门，然后边用，边查资料，边提高。

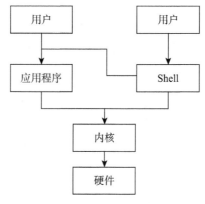

图 3-1　Shell 架构

Shell 脚本实例（来源：/qnn/qcom/aistack/qnn/******/examples/QNN/NetRun/android）：

\# 省略

MODEL=qnn_model_8bit_quantized

adb push ${QNN_LIB_ROOT}/libQnnHtp.so ${TARGET_ROOT}

adb wait-for-device push ${QNN_LIB_ROOT}/libQnnHtpPrepare.so ${TARGET_ROOT}

adb wait-for-device push ${QNN_LIB_ROOT}/libQnnHtpV69Stub.so ${TARGET_ROOT}

adb wait-for-device push ${HEXAGON_V69_SKEL_PATH} ${TARGET_ROOT}

ADSP_LIBRARY_PATH=" export
　　　　　　　ADSP_LIBRARY_PATH=\" " ${TARGET_ROOT}":
　　　　　　　/vendor/dsp/cdsp:/vendor/lib/rfsa/adsp:/system/lib/rfsa/
　　　　　　　adsp:/dsp\" "
LD_LIBRARY_PATH=" export
　　　　　　　LD_LIBRARY_PATH=" ${TARGET_ROOT}" :/vendor/
　　　　　　　dsp/cdsp:/vendor/lib64/:\$LD_LIBRARY_PATH"
QNN_NET_RUN_CMD=" ${ADSP_LIBRARY_PATH} && ${LD_LIBRARY_PATH}
　　　　　　　&& ./qnn-net-run --model lib" ${MODEL}" .so
　　　　　　　--input_list input_list_float.txt
　　　　　　　--backend libQnnDspV65Stub.so"

echo" INFO: Executing qnn-net-run."
ADB_SHELL_CMD=" export
　　　　　　　LD_LIBRARY_PATH=/data/local/tmp/qnn/" ${MODEL}"
　　　　　　　&& cd
　　　　　　　/data/local/tmp/qnn/" ${MODEL}" &&" ${QNN_NET_
　　　　　　　RUN_CMD}

adb shell ${ADB_SHELL_CMD}
省略
```

安装 Clang 的 Shell 脚本（来源：/qnn/qcom/aistack/qnn/\*\*\*\*\*\*/bin/check-python-dependency）：

```
#!/bin/bash

function setup_clang_9 ( ) {
  sudo apt-get update
  pkgs_to_check= (' clang-9' ' libc++-9-dev')
  j=0
  while [$j -lt ${#pkgs_to_check [*] }]; do
    install_status=$ (verify_pkg_installed ${pkgs_to_check [$j] })
    if [" $install_status" ==" "]; then
      sudo apt-get install -y ${pkgs_to_check [$j] }
      if [[$? -ne 0]]; then
        echo" ERROR: Failed to install required packages for clang9"
```
```

```
    fi
  fi
  j=$ (( $j +1 ) );
 done
}

setup_clang_9
echo" Done!!"
```

安装之后,进行验证,验证方法如下:

```
$ clang --version
clang version 10.0.0-4ubuntu1
Target: x86_64-pc-linux-gnu
Thread model: posix
InstalledDir: /usr/bin
```

3.2 环境变量配置

Linux 环境变量的配置方法有三种,分别为 bashrc、profile 和 environment。

它们的区别如表 3-1 所示,其中 LIBRARY_PATH 和 LD_LIBRARY_PATH 是 Linux 中两个环境变量,其区别是前者是编译 GCC 时用到,后者是运行加载时用于指定查找共享库(动态链接库)除默认路径(/lib、/usr/lib)之外的其他路径。比如我们在 ARM 终端中测试如下:

root@TinaLinux:/# echo $LD_LIBRARY_PATH
./libcv:/usr/lib/eyesee-mpp:/usr/lib/eyesee-mpp:
root@TinaLinux:/# echo $LIBRARY_PATH

表 3-1 Linux 环境变量设置和生效方法

修改 PATH 查看 PATH:echo $PATH	生效方法	有效期限	用户权限
gedit ~/.bashrc 在最后一行添加: export PATH=" /home/user/anaconda3/bin:$PATH" export PATH=$PATH:/usr/local/cuda/bin export LD_LIBRARY_PATH=/usr/local/cuda-9.0/bin/lib64:$LD_LIBRARY_PATH	关闭当前终端,重新打开一个新终端或 source ~/.bashrc	永久有效	仅当前用户

续表

修改 PATH 查看 PATH：echo $PATH	生效方法	有效期限	用户权限
gedit /etc/profile 添加 export PATH=/usr/local/bin:$PATH	sudo reboot	永久有效	所有用户
gedit /etc/environment 在 PATH="/usr/local/bin:/usr/sbin" 中加入 :/usr/local/bin	sudo reboot	永久有效	所有用户
export PATH=/usr/local/bin:$PATH echo $PATH 查看	立即生效	临时改变，仅在当前终端有效	仅当前用户

3.3 编译器 GCC

GCC（GNU Compiler Collection，GNU 编译器套件）是 GNU 开发的用于编译 C、C++、Objective-C、Fortran、Ada、Go 等编程语言的编译器，它包括 GCC 和 G++ 工具，均可以用来编译 C 和 C++ 程序。如图 3-2 所示，G++ 是 GNU 组织开发的 C++ 编译器。

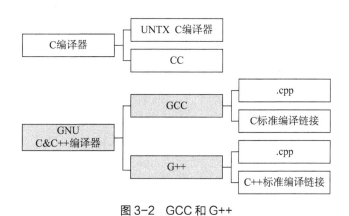

图 3-2　GCC 和 G++

我们可以通过以下命令查询图 3-2 中的工具安装路径：

（base）user@ubuntu:~/cv/cvtest$ which cc

/usr/bin/cc

（base）user@ubuntu:~$ which g++

/usr/bin/g++

（base）user@ubuntu:~$ which gcc

/usr/bin/gcc

（base）user@ubuntu:~$

其中 G++ 的使用语法格式如下：

g++［选项］编译文件［选项］［目标文件］

以下是简单的实现 C&C++ 代码的编译实例：

（base）user@ubuntu:~/cv/cvtest$ g++ \`pkg-config opencv4--cflags\` test.cpp-o test_g++ \`pkg-config opencv4--libs\` -std=c++11-O3-fno-asm-v-fPIC

-o test_g++ 表示编译生成目标文件名称；-O3 表示编译器的优化级别（-O0 表示没有优化，-O1 为缺省值，-O3 优化级别最高）；-fPIC 表明编译后生成地址无关代码（Position-independent Code，PIC）；\`pkg-config opencv4--cflags\` 和 \`pkg-config opencv4--libs\`，具体代表的意义如下所示：

（base）user@ubuntu:~/cv/cvtest$ \`pkg-config opencv4--cflags\`

bash: -I/usr/local/include/opencv4: No such file or directory

（base）user@ubuntu: ~/cv/cvtest$

（base）user@ubuntu: ~/cv/cvtest$ \`pkg-config opencv4--libs\`

bash: -L/usr/local/lib: No such file or directory

（base）user@ubuntu: ~/cv/cvtest$

（base）user@ubuntu: ~/cv/cvtest/opencv/build$ pkg-config--cflags opencv4

-I/usr/local/include/opencv4

（base）user@ubuntu: ~/cv/cvtest/opencv/build$ pkg-config--libs opencv4

-L/usr/local/lib-lopencv_stitching-lopencv_ml-lopencv_gapi-lopencv_video-lopencv_photo-lopencv_dnn-lopencv_objdetect-lopencv_highgui-lopencv_videoio-lopencv_imgcodecs-lopencv_calib3d-lopencv_features2d-lopencv_imgproc-lopencv_flann-lopencv_core

3.4　CMake 项目构建、Make 编译

GCC 编译指令一般对较小项目工程操作很方便，但是对于一个大型项目工程，由于源文件数量很多、类型不同、依赖关系复杂，如果直接用 GCC 命令编译就很复杂，而且容易混乱。Make 是一个自动化编译工具，Make 命令本身没有编译和链接的功能，它依据 Makefile 文件中命令进行编译和链接（Makefile 文件中指定 gcc 等其他编译器），大大提高编译效率。

Makefile 文件可以人工手写，但是比较麻烦，于是 CMake 工具出场了。通过 CMake 命令将 CMakeLists.txt 文件（如图 3-3 所示）转化成为 Make 所需要的 Makefile 文件，实现项目构建，再用 Make 命令进行编译。其中 CMakeList.txt 文件由命令、注释和空格组成，它的语法比较简单，较容易编写。[①]

---

① 更多资料可以参考：https://cmake.org/cmake/help/latest/guide/tutorial/index.html

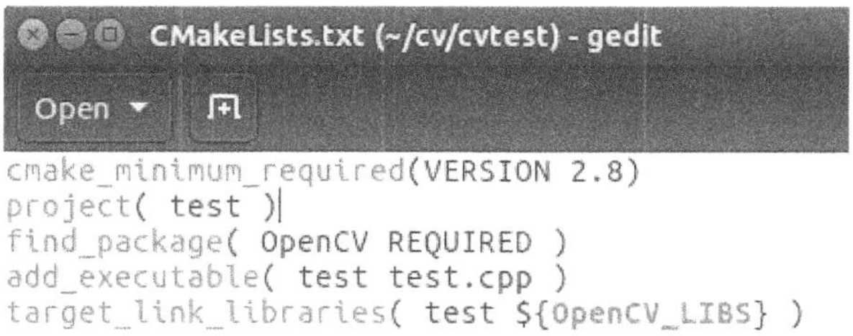
CMakeLists.txt (~/cv/cvtest) - gedit
Open
cmake_minimum_required(VERSION 2.8)
project(test)
find_package(OpenCV REQUIRED)
add_executable(test test.cpp)
target_link_libraries(test ${OpenCV_LIBS})

图 3-3　CMakeLists.txt 文件

以上讨论的方法具体的操作流程如下：

```
user@ubuntu:~/cv/cvtest$ cmake .
-- The C compiler identification is GNU 5.4.0
-- The CXX compiler identification is GNU 5.4.0
-- Check for working C compiler: /usr/bin/cc
-- Check for working C compiler: /usr/bin/cc -- works
-- Detecting C compiler ABI info
-- Detecting C compiler ABI info - done
-- Detecting C compile features
-- Detecting C compile features - done
-- Check for working CXX compiler: /usr/bin/c++
-- Check for working CXX compiler: /usr/bin/c++ -- works
-- Detecting CXX compiler ABI info
-- Detecting CXX compiler ABI info - done
-- Detecting CXX compile features
-- Detecting CXX compile features - done
-- Found OpenCV: /usr/local ( found version" 4.5.2" )
CMake Warning ( dev ) at CMakeLists.txt:4 ( add_executable ) :
  Policy CMP0037 is not set: Target names should not be reserved and should
  match a validity pattern. Run" cmake--help-policy CMP0037" for policy
  details.  Use the cmake_policy command to set the policy and suppress this
  warning.

  The target name" test" is reserved or not valid for certain CMake features,
  such as generator expressions, and may result in undefined behavior.
This warning is for project developers. Use-Wno-dev to suppress it.

-- Configuring done
```

```
-- Generating done
-- Build files have been written to: /home/user/cv/cvtest
user@ubuntu: ~/cv/cvtest$

user@ubuntu: ~/cv/cvtest$ make
Scanning dependencies of target test
[ 50%] Building CXX object CMakeFiles/test.dir/test.cpp.o
[100%] Linking CXX executable test
[100%] Built target test
user@ubuntu: ~/cv/cvtest$
```

关于 CMakeLists.txt 文件，我们先要能够看懂和可以修改，然后才是掌握如何编写。读者朋友可以在项目实践中结合手册或网站资料一步步去学和用。

3.5　LLVM 和 Clang

LLVM 是模块化和可重用的编译器和工具链技术的集合（This is the LLVM organization on GitHub for the LLVM Project: a collection of modular and reusable compiler and toolchain technologies.[①]）。如上一节中的 GCC 为传统编译器，它的流程分为三个阶段：前端（Frontend）、优化器（Optimizer）和后端（Backend），但是三个阶段中有耦合，如图 3-4 所示。

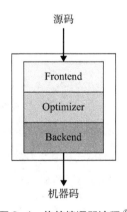

图 3-4　传统编译器流程[②]

如果其中的优化器阶段使用语言无关的中间代码表示，那么带来的好处是实现编译器三个阶段高度模块化，编译器更加容易扩展，可以做到：支持多种编程语言

---

① 引文摘自 https://github.com/llvm
② 图参考：http://www.aosabook.org/en/llvm.html

时，只需添加多个 Frontend；支持多种目标处理器，只需添加多个 Backend。LLVM IR（Intermediate Representation，中间端表达式）是一种优化器阶段统一的中间代码，如图 3-5 所示。

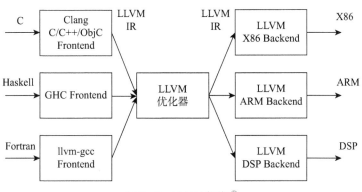

图 3-5　LLVM 架构[①]

Clang 是 LLVM 项目的一个子项目，是 LLVM 的 C/C++/Objective-C 编译器前端。Clang 可以实现快速编译，提供非常有用的错误和警告消息，其中 Clang Static Analyzer 和 Clang tidy 工具可以自动查找代码中的错误。

3.6　Android SDK 和 NDK 配置

讲述以下内容之前，读者朋友需要先明白 Android SDK 和 NDK 是不一样的，不能搞混淆，现在进入正文。

（1）Android SDK 配置

从 http://tools.android-studio.org/index.php/sdk/ 下载 Android SDK（本书以 android-sdk_r24.4.1-linux.tgz 版本为示例），解压到指定文件夹，改名为 android-sdk-linux。配置环境变量，如下所示。

$ export ANDROID_HOME=./android-sdk-linux
$ export PATH=$ANDROID_HOME/tools:$PATH
$ source /etc/profile

按照以下命令打开 SDK Manager 验证是否配置成功。如图 3-6 所示，表明配置成功。

$ android

---

① 图参考：http://www.aosabook.org/en/llvm.html

图 3-6　Android SDK Manager

（2）Android NDK 配置

Android NDK（Native Development Kit）是 Android 的开发和编译工具包，从 https://developer.android.google.cn/ndk/downloads 下载 Android NDK，如图 3-7 所示。

图 3-7　Android NDK

配置 NDK 如下：

$ export ANDROID_NDK=******/android-ndk-r25b

$ export PATH=$PATH:$ANDROID_NDK

$ source /etc/profile

接下来运行以下命令验证是否配置成功。

$ ndk-build

Android NDK: Could not find application project directory!

Android NDK: Please define the NDK_PROJECT_PATH variable to point to it.

******/android-ndk-r25b/build/core/build-local.mk:151: *** Android NDK: Aborting   . Stop.

以上打印的日志，表明已经配置成功。

下面介绍一下 CMake 的配置，在项目开发中会用到。

（3）CMake 配置

export CMAKE=******/cmake-3.23.1-linux-x86_64/bin/cmake

3.7　Clang/Clang++ 编译 C/C++/Neon 程序和终端运行

如何编译一个用 C/C++/Neon 语言编写的程序呢？又如何实现在终端设备中运行测试呢？这是接下来所要讨论的内容。我们知道自 Android NDK r13b 版本起，LLVM/Clang 成为 Android NDK 的默认工具链（NDK_TOOLCHAIN_VERSION now defaults to Clang.），从 Android NDK r18b 版本起已经删除 GCC，所以 Clang 成为主流编译工具。如图 3-8 所示是 Android NDK 的文件结构。

```
android-ndk-r25b$ ls
build           ndk-gdb        NOTICE            README.md        sources
CHANGELOG.md    ndk-lldb       NOTICE.toolchain  shader-tools     toolchains
meta            ndk-stack      prebuilt          simpleperf       wrap.sh
ndk-build       ndk-which      python-packages   source.properties
```

图 3-8　Android NDK 的文件结构

我们进入如下路径，用 tree 命令查看一下有哪些编译工具。

android-ndk-r25b/toolchains/llvm/prebuilt/linux-x86_64/bin$ tree
.
├── aarch64-linux-android21-clang
├── aarch64-linux-android21-clang++
……
├── bisect_driver.py
├── clang -> clang-14
├── clang++ -> clang
├── clang-14
├── clang-check
├── clangd
├── clang-format
├── clang-tidy
├── dsymutil
├── git-clang-format
├── i686-linux-android19-clang
├── i686-linux-android19-clang++
……

```
├── ld -> ld.lld
├── ld64.lld -> lld
├── ld.lld -> lld
├── lld
├── lldb
├── lldb-argdumper
├── lldb.sh
├── lld-link -> lld
├── llvm-addr2line -> llvm-symbolizer
├── llvm-ar
├── llvm-as
├── llvm-bolt
├── llvm-cfi-verify
├── llvm-config
├── llvm-cov
├── llvm-cxxfilt
├── llvm-dis
├── llvm-dwarfdump
├── llvm-dwp
├── llvm-lib -> llvm-ar
├── llvm-link
├── llvm-lipo
├── llvm-modextract
├── llvm-nm
├── llvm-objcopy
├── llvm-objdump
├── llvm-profdata
├── llvm-ranlib -> llvm-ar
├── llvm-rc
├── llvm-readelf -> llvm-readobj
├── llvm-readobj
├── llvm-size
├── llvm-strings
├── llvm-strip -> llvm-objcopy
```

```
├── llvm-symbolizer
├── llvm-windres -> llvm-rc
├── merge-fdata
├── remote_toolchain_inputs
├── sancov
├── sanstats
├── scan-build
├── scan-view
├── x86_64-linux-android21-clang
├── x86_64-linux-android21-clang++
……
└── yasm
```

接下来按照以下命令配置编译器环境以及查看编译器的版本。

$ export ANDROID_NDK_BIN=******/android_ndk_test/android-ndk-r25b/toolchains/llvm/prebuilt/linux-x86_64/bin

$ ${ANDROID_NDK_BIN}/aarch64-linux-android21-clang++ -v

Android（******，based on ******）clang version 14.0.6（https://android.googlesource.com/toolchain/llvm-project ******）

Target: aarch64-unknown-linux-android21

Thread model: posix

InstalledDir: ******/android-ndk-r25b/toolchains/llvm/prebuilt/linux-x86_64/bin

再按照以下命令编译一个简单的测试程序，其程序代码如下所示。

#include <iostream>
using namespace std;
#include <arm_neon.h> // #include <arm_neon.h> 文件包含编译 Neon 程序的依赖文件。

int main（）
{
 cout <<″computer vision″<< endl；
 return 0；
}

执行以下命令编译这个测试程序。

$ ${ANDROID_NDK_BIN}/aarch64-linux-android21-clang++ test.cpp -o test

$ ls

back test test.cpp

以上测试程序编译成功之后，我们得到一个可执行文件 test。通过 ADB 工具实现在安卓设备中运行。

adb push test data/

进入智能设备终端中，直接运行 ./test 执行可执行文件。（如果运行时发现没有运行权限，我们需要运行一下命令 chmod 777 test）

通过以上的内容，我们现在已经掌握了如何基于手机 CPU 处理器开发 C/C++/Neon 语言程序了。

3.8　GDB 调试

GDB（GNU project debugger）工具是一种基于 Linux 系统的代码调试工具，主要用于 C/C++ 代码开发。借助 GDB 可调试定位程序的缺陷（Bug）。由于没有图形调试界面，所以使用较麻烦。

调试的程序在 G++ 等工具编译时要添加 -g 配置，用以产生调试信息，才能支持调试，具体编译方法如下所示。

（base）user@ubuntu:~/cv/cvtest$ g++ `pkg-config opencv4 --cflags` facedetect.cpp -o facedetect_g++_O3 `pkg-config opencv4 --libs` -std=c++11 -fPIC -g -O3

图 3-9 所示为 GDB 的调试使用过程，是笔者在实际项目中的调试经历。更多的使用方法，建议读者朋友参考学习有关 GDB 调试方面的书籍，考虑到篇幅的限制，笔者在此不做过多的讲解。总结一句话：GDB 调试对于小型程序好用，复杂的程序建议使用可视化界面调试工具。

（a）启动 GDB 调试

```
(gdb) l
4       #include "opencv2/videoio.hpp"
5       #include <iostream>
6
7       using namespace std;
8       using namespace cv;
9
10
11      void detectAndDraw( Mat& img, CascadeClassifier& cascade,
12                          double scale);
13
(gdb)
14      string cascadeName;
15
16
17      int main( int argc, const char** argv )
18      {
19
20          Mat frame, image;
21          string inputName;
22
23          CascadeClassifier cascade;
(gdb)
```

(b) GDB 调试，l 命令查看源码

```
(gdb) b 49
Breakpoint 1 at 0x401641: file facedetect.cpp, line 49.
(gdb) r
Starting program: /home/user/cv/cvtest/facedetect_g++_O3
[Thread debugging using libthread_db enabled]
Using host libthread_db library "/lib/x86_64-linux-gnu/libthread_db.so.1".
Detecting face(s) in lena.jpg

Breakpoint 1, main (argc=<optimized out>, argv=<optimized out>)
    at facedetect.cpp:49
49              detectAndDraw( image, cascade, scale);
(gdb)
```

(c) GDB 调试中设置断点和运行程序，b 文件名：行号

```
(gdb) s
detectAndDraw (img=..., cascade=..., scale=scale@entry=1) at facedetect.cpp:58
58      {
(gdb) n
71          };
(gdb) n
60          vector<Rect> faces, faces2;
(gdb) n
71          };
(gdb) n
72          Mat gray, smallImg;
(gdb) n
74          cvtColor( img, gray, COLOR_BGR2GRAY );
(gdb) n
[New Thread 0x7ffff4ad1700 (LWP 42402)]
[New Thread 0x7fffefff700 (LWP 42403)]
[New Thread 0x7fffe7fff700 (LWP 42404)]
75          double fx = 1 / scale;
(gdb) n
76          resize( gray, smallImg, Size(), fx, fx, INTER_LINEAR_EXACT );
(gdb) p fx
$1 = <optimized out>
(gdb) n
75          double fx = 1 / scale;
(gdb) n
76          resize( gray, smallImg, Size(), fx, fx, INTER_LINEAR_EXACT );
(gdb) p fx
$2 = 1
(gdb)
```

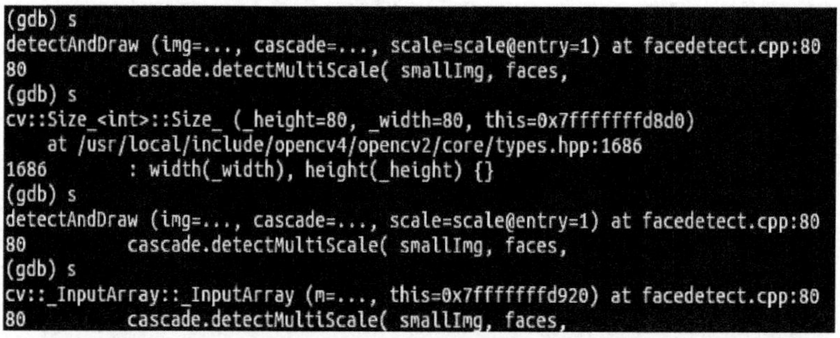

```
(gdb) s
detectAndDraw (img=..., cascade=..., scale=scale@entry=1) at facedetect.cpp:80
80              cascade.detectMultiScale( smallImg, faces,
(gdb) s
cv::Size_<int>::Size_ (_height=80, _width=80, this=0x7fffffffd8d0)
    at /usr/local/include/opencv4/opencv2/core/types.hpp:1686
1686        : width(_width), height(_height) {}
(gdb) s
detectAndDraw (img=..., cascade=..., scale=scale@entry=1) at facedetect.cpp:80
80              cascade.detectMultiScale( smallImg, faces,
(gdb) s
cv::_InputArray::_InputArray (m=..., this=0x7fffffffd920) at facedetect.cpp:80
80              cascade.detectMultiScale( smallImg, faces,
```

（d）GDB 调试中，n 命令实现单语句执行，s 命令实现单步跟踪到函数内部

图 3-9 使用 GDB 进行调试

对于使用 C/C++ 算法的开发调试工具，笔者建议使用 Visual Studio（VS）或 Visual Studio Code（VS Code），毕竟 GDB 命令行模式用于调试还不是很方便。我们先用 IDE 工具把算法模块调试验证好，再集成到目标平台，往往会事半功倍。

3.9 ADB 工具

ADB（Android Debug Bridge，Android 调试桥）是一个终端调试命令行工具。它可以实现 PC 端操作 Android 终端设备，具体功能包括：

a. 运行终端设备的 Shell 命令；

b. PC 和终端设备之间上传和下载文件；

c. 管理设备端口映射；

d. 安装和卸载终端设备 APK 程序。

首先运行命令 sudo apt-get install android-tools-adb 安装 ADB 工具。

安装完成后，运行以下命令可以查看 ADB 的版本号信息。如果有具体的版本信息输出，说明已经安装成功。

```
$ adb version
Android Debug Bridge version ******
Version ******
Installed as /usr/lib/android-sdk/platform-tools/adb
```

现在可以运行以下命令列出所有可以访问到的终端设备，其中 id1、id2 代表设备号。

```
(base)user@user:~$ adb devices
List of devices attached
id1  device
id2  device
```

运行以下命令进入指定设备号的设备以及运行更多 Shell 命令。

abd-s 设备号 shell

一些常用的命令介绍如下：

实现本地电脑文件复制到目标设备：

adb-s ****** push 本地路径 设备路径

实现设备文件复制到本地电脑：

adb-s ****** pull 设备路径 本地路径

abd-s 设备号 shell mkdir（ls、pwd、cp、rm 等更多命令）

如果在执行命令 adb devices 后，发现打印结果只有一行"List of devices attached"，这表明找不到设备。遇到这个问题，首先通过 lsusb 命令查看 PC 机器的 USB 设备列表中是否有这个终端设备。

$ lsusb

Bus 002 Device 002: ID ……:…… Intel Corp.

Bus 002 Device 001: ID ……:…… Linux Foundation 2.0 root hub

Bus 006 Device 001: ID ……:…… Linux Foundation 3.0 root hub

Bus 005 Device 001: ID ……:…… Linux Foundation 2.0 root hub

Bus 003 Device 001: ID 1……:0…… Linux Foundation 2.0 root hub

如果 USB 设备比较多，可以先拔掉 USB 数据线，再运行一下 lsusb 命令查看当前 USB 设备列表，前后对比观察看是否能找到这个设备。如果能够找到这个设备，可以排除是设备驱动或硬件相关的问题，反之亦然。然后运行命令 adb kill-sever、adb start-server 实现先关闭 ADB 服务，再重新启动 ADB 服务。如果还是不能找到设备，可以尝试以下的配置修改方法。

(base) user@user:~$ sudo vi /etc/udev/rules.d/51-android.rules

SUBSYSTEM=="usb", ATTR{idVendor}=="1……", ATTR{idProduct}=="0……", MODE="0600", OWNER=="user"

(base) user@user:~$ sudo chmod a+r /etc/udev/rules.d/51-android.rules

(base) user@user:~$ sudo vi ~/.android/adb_usb.ini

0x******

(base) user@user:~$ sudo adb kill-server

(base) user@user:~$ sudo adb devices

List of devices attached

* daemon not running; starting now at tcp:5037

* daemon started successfully

****** device

通过以上方法，发现已经可以找到设备，接下来我们进入设备测试，如图 3-10 所示。

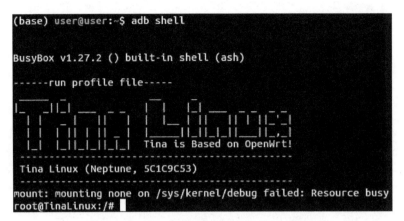

图 3-10　运行 adb shell 命令进入 ARM 终端

ADB 工具中使用 $ adb logcat 抓取终端设备运行的日志，结合"adb logcat | grep 关键词"进行运行程序的问题定位。在使用前，先使用 adb logcat-c 清空日志缓存。

第 4 章

算子和图优化

算子是什么？

深度学习中的模型是由一系列的计算小模块组成的，我们把这些小模块称为算子，英文名称为 Operator，简称 OP。更加宽泛的定义：算子是实现一种计算逻辑的模块，不仅针对深度学习模型，也可以是机器学习模型。

图又是什么呢？

图是深度学习中由一系列算子组成的一个有向无环图（Directed Acyclic Graph，DAG）。图源于离散数学中的图论（Graph Theory），有向无环图是从图中的任一节点沿着边的方向遍历后不能返回到该节点的图，也就是图中没有环。基于图提出的图卷积网络（Graph Convolution Network，GCN）给深度学习带来很多新的惊喜。读者朋友可以参考相关的图论书籍做进一步的研究，相信对深度学习模型的开发会有新的发现。按照计算逻辑颗粒度划分，图可以分为一个个子图，然后构成一个完整的大图，这个大图就是深度学习模型。

总结一下：算子构成子图，子图构成模型。

算子和图的性能直接关系到模型的推理性能，算子和图的优化非常关键。本章关于算子和图的优化方法主要包括以下：

（1）算法层优化，如通过 Img2col+GEMM 或 Winograd 优化卷积计算；

（2）硬件层优化（处理器硬件资源优化），如 SIMD 指令向量化、CUDA 并行优化、优化内存访问（充分利用寄存器和高速缓存）、参数共享、AI 编译器实现的图优化等。

通过以上方法我们可以实现 Conv、MaxPool、AveragePool、Max、Add、Pow、Sigmoid、Softmax、ReLU、Expand、Dropout、Concat、BatchNormalization、Unsqueeze、Upsample、Resize、LSTM、PixelShuffle 等算子在目标硬件中的性能以及这些算子之间的融合优化。

4.1 算法层优化

算法层（Algorithm layer）优化是从算法的数学计算原理出发，简化计算时间复杂度或者空间复杂度的过程。如下文讨论的 Img2col+GEMM 和 Winograd 卷积优化。因为影响算法计算时间的开销因素有计算和访存两方面，因此可以将算子分为以下两类：

（1）计算密集型算子，如 Conv、GEMM；

（2）访存密集型算子，如 ReLU、BN、Concat、PixelShuffle（上采样模块，同 depth_to_space）、space_to_depth（下采样）、Transpose、Reshape 等。

4.1.1　Img2col+GEMM 优化卷积

深度学习模型中基本的算子有卷积层［Convolution，图 4-1（a）］、池化层［Pooling，图 4-1（b）］和激活层［Activation，图 4-1（c）］以及全连接层。

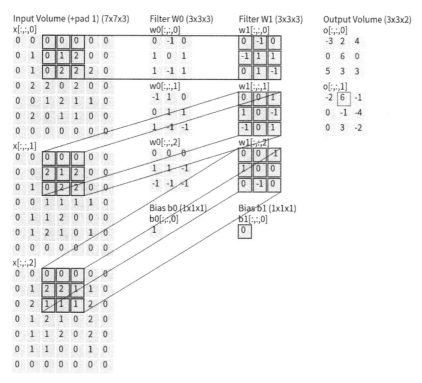

(a)标准卷积层计算原理

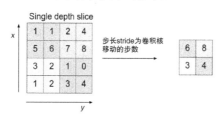

(b)池化层(步长 stride 为卷积核移动的步数)

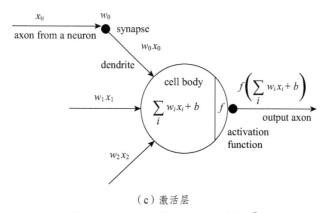

(c)激活层

图4-1 深度学习模型中基本的算子[1]

---

[1] 图来源:CS231n Convolutional Neural Networks for Visual Recognition,https://cs231n.github.io/convolutional-networks/#overview

图4-1（a）中W0、W1代表2个3×3大小、3通道（Channels）的卷积核。卷积核在卷积计算过程中每次移动2个步长（先水平方向，后向下方向）。卷积刚开始的输入通道取决于图片的类型。如果是RGB图片，则Channels等于3。图中的输入维度是5×5×3，填充（Padding）后维度是7×7×3。标准卷积计算完成后输出的特征层（Feature Map）的通道数等于参加计算的卷积核的个数，图4-1（a）所示是2个卷积核，所以卷积计算后得到2个特征层，也就是输出通道为2。

以上讨论的是卷积计算中的输入通道数和输出通道数。

图4-1（a）展示的标准卷积采用滑窗的方式进行计算，这样很容易理解，但是在实际计算中的计算量很大，所以不能直接应用在推理和训练中，需要研究一种快速卷积算法来实现卷积计算。目前比较常用的方法有Img2col+GEMM、Winograd算法和FFT（Fast Fourier Transforms）变换法（NNPACK中卷积采用FFT+Winograd结合）等。

接下来我们先讨论基于Img2col+GEMM法加速卷积计算，其计算示意如图4-2所示。

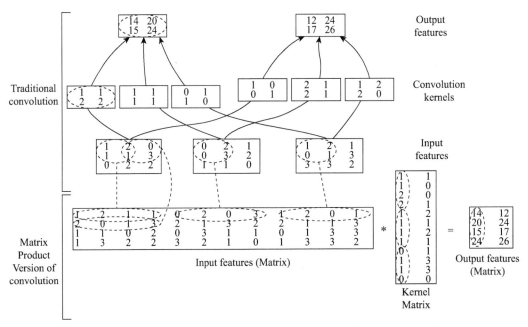

图4-2 Img2col+GEMM卷积运算[①]

为了更容易理解上图，我们可以把图4-2再做分解，首先输入张量，如图4-3所示，维度为3×3×3。

---

① 图来源：High Performance Convolutional Neural Networks for Document Processing Kumar Chellapilla, Sidd Puri, Patrice Simard

1	2	0
1	1	3
0	2	2

0	2	1
0	3	2
1	1	0

1	2	1
0	1	3
3	3	2

图 4-3　输入张量

然后有 2 个卷积核，每个卷积核有 3 个通道，如图 4-4 所示。

1	1		1	0
2	2		0	1

1	1		2	1
1	1		2	1

0	1		1	2
1	0		2	0

图 4-4　2 个卷积核

我们将标准卷积计算过程中的每次滑窗区域 mask（掩码）的矩阵展平得到一个个行向量，然后组合成为一个矩阵，具体流程如图 4-3 所示。

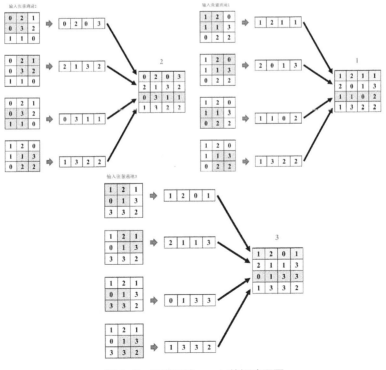

图 4-5　滑窗区域 mask 的矩阵展平

再将上图中的三个矩阵按照行扩展方向合并为一个大矩阵,如图 4-6 所示。

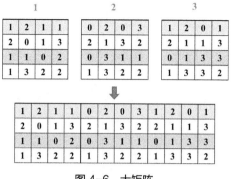

图 4-6　大矩阵

此外还需要将卷积核中的 3 个通道分别展平得到列向量,然后合并成为一个列向量,如图 4-7 所示。

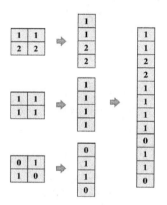

图 4-7　列向量

现在将图 4-6 的大矩阵和图 4-7 的列向量做矩阵乘法运算,把结果再 reshape(矩阵形状改变)一下,就得到卷积的结果,如图 4-8 所示。

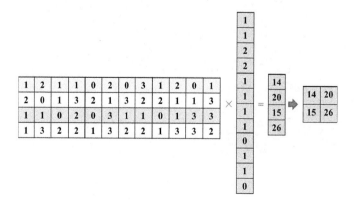

图 4-8　矩阵乘法运算

以上的运算涉及矩阵计算，比如通用矩阵乘法（General Matrix Multiplication，GEMM）。矩阵计算在 BLAS 库、Intel 的 MKL、AMD 的 ACML、NVIDIA 的 cuBLAS、开源的 OpenBLAS，均针对特定的处理器进行过针对性的优化加速。

Img2col+GEMM 优化卷积的原理是用空间换时间，其缺点为需要消耗更多的内存资源。特别是当滑窗步长小于卷积核大小时，会导致转换后的大矩阵有大量的重复值，会更加浪费内存资源。针对这个问题，研究者提出了进一步的优化方案，如图 4-9 所示的扩展有限列算法。鉴于篇幅限制，笔者在此不做过多的讨论。

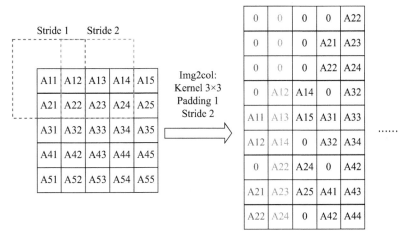

图 4-9　部分 Img2col 扩展有限列（其中 kernel 大小为 3x3，padding=1，stride=2）[1]

在 CMSIS、Tengine、caffe 等框架中的卷积算子一般采用 Img2col+GEMM 的优化方法，它们的实现代码如下。

代码出自：https://github.com/pjreddie/darknet/blob/master/src/im2col.c

```c
#include "im2col.h"
#include <stdio.h>
float im2col_get_pixel(float *im, int height, int width, int channels, int row, int col, int channel, int pad)
{
    row -= pad;
    col -= pad;

    if(row < 0 || col < 0 || row >= height || col >= width)
        return 0;
```

---

[1] 图来源 Lai L, Suda N, Chandra V. CMSIS-NN: Efficient Neural Network Kernels for Arm Cortex-M CPUs. 2018. DOI:10.48550/arXiv.1801.06601

```
    return im [ col + width* ( row + height*channel ) ];
}

//From Berkeley Vision's Caffe!
//https://github.com/BVLC/caffe/blob/master/LICENSE
void im2col_cpu ( float* data_im, int channels, int height, int width, int ksize, int stride, int pad, float* data_col )
{
    int c, h, w;
    int height_col = ( height + 2*pad - ksize )/ stride + 1;
    int width_col = ( width + 2*pad - ksize )/ stride + 1;

    int channels_col = channels * ksize * ksize;
    for ( c = 0; c < channels_col; ++c ) {
        int w_offset = c % ksize;
        int h_offset = ( c / ksize ) % ksize;
        int c_im = c / ksize / ksize;
        for ( h = 0; h < height_col; ++h ) {
            for ( w = 0; w < width_col; ++w ) {
                int im_row = h_offset + h * stride;
                int im_col = w_offset + w * stride;
                int col_index = ( c * height_col + h )* width_col + w;
                data_col [ col_index ]= im2col_get_pixel ( data_im, height, width, channels, im_row, im_col, c_im, pad );
            }
        }
    }
}
```

4.1.2 Winograd 优化卷积

上一节中讨论的 Img2col+GEMM 方法用于加速卷积计算在工程中具有很高的实用性。本节我们将讨论另一种更加主流的加速方法——Winograd，它广泛应用在 TNN、NCNN、Paddlelite、FeatherCNN、NNPACK、MNN、MACE、ARM-ComputeLibrary 等推理框架中。Winograd 的思想是通过减少比加法更加耗时的乘法的次数来实现卷积计算加速。现在我们讨论一下 Winograd 的实现原理。

首先把输入数据（以一维数据为例，二维数据同理）定义为：

$d = \begin{bmatrix} d_0 & d_1 & d_2 & d_3 \end{bmatrix}^T$

卷积定义为：

$g = \begin{bmatrix} g_0 & g_1 & g_2 \end{bmatrix}^T$

卷积计算中滑窗移动得到的矩阵为：

$$\begin{bmatrix} d_0 & d_1 & d_2 \\ d_1 & d_2 & d_3 \end{bmatrix}$$

关于以上这个矩阵，我们可以发现其中有规律地分布着大量的重复元素，如第 1 行和第 2 行中均有 d_1 和 d_2。

那么优化的计算方式可以表示为

$$F(2, 3) = \begin{bmatrix} d_0 & d_1 & d_2 \\ d_1 & d_2 & d_3 \end{bmatrix} \begin{bmatrix} g_0 \\ g_1 \\ g_2 \end{bmatrix} = \begin{bmatrix} m_1 + m_2 + m_3 \\ m_2 - m_3 - m_4 \end{bmatrix} \quad (4.1\text{-}1)$$

结果矩阵中的 m_1、m_2、m_3、m_4 值如下：

$$m_1 = (d_0 - d_2) g_0 \quad m_2 = (d_1 + d_2) \frac{g_0 + g_1 + g_2}{2}$$
$$m_4 = (d_1 - d_3) g_2 \quad m_3 = (d_2 - d_1) \frac{g_0 - g_1 + g_2}{2} \quad (4.1\text{-}2)$$

$$F(2, 3) = \begin{bmatrix} d_0 & d_1 & d_2 \\ d_1 & d_2 & d_3 \end{bmatrix} \begin{bmatrix} g_0 \\ g_1 \\ g_2 \end{bmatrix} = \begin{bmatrix} r_0 \\ r_1 \end{bmatrix} \quad (4.1\text{-}3)$$

$$Y = A^T \left[(Gg) \odot (B^T d) \right] \quad (4.1\text{-}4)$$

$$B^T = \begin{bmatrix} 1 & 0 & -1 & 0 \\ 0 & 1 & 1 & 0 \\ 0 & -1 & 1 & 0 \\ 0 & 1 & 0 & -1 \end{bmatrix} \quad (4.1\text{-}5)$$

$$G = \begin{bmatrix} 1 & 0 & 0 \\ \frac{1}{2} & \frac{1}{2} & \frac{1}{2} \\ \frac{1}{2} & -\frac{1}{2} & \frac{1}{2} \\ 0 & 0 & 1 \end{bmatrix} \quad (4.1\text{-}6)$$

$$A^T = \begin{bmatrix} 1 & 1 & 1 & 0 \\ 0 & 1 & -1 & -1 \end{bmatrix} \quad (4.1\text{-}7)$$

$$d = Y = A^T [[GgG^T] \odot [B^T dB]] A [d_0 \quad d_1 \quad d_2 \quad d_3]^T \quad (4.1\text{-}8)$$

这样的好处是不需要计算点积就可以实现卷积运算。公式中的 $\frac{g_0 + g_1 + g_2}{2}$ 和 $\frac{g_0 - g_1 - g_2}{2}$ 也不需要每一次卷积计算都计算一次，因为 g_0、g_1、g_2 是来自卷积核的值，是固定的参数，对于模型训练阶段在卷积计算之前只需要计算一次就可以实现复用，对于模型推理阶段在推理之前计算并且保存这两个值，就可以实现推理中一直复用。

以上公式中，我们发现只需要进行 4 次加法和 4 次乘法计算，就可以得到 m_1、m_2、m_3、m_4 的值，然后使用 4 个加法操作得到结果。而普通的点积计算如下：

$r_0 = (d_0 \cdot g_0) + (d_1 \cdot g_1) + (d_2 \cdot g_2)$

$r_1 = (d_1 \cdot g_0) + (d_2 \cdot g_1) + (d_3 \cdot g_2)$

需要 6 次乘法计算。Winograd 实现将乘法计算减少了 $\frac{1}{3}$，因为乘法比加法更加消耗处理器资源，所以对卷积计算的优化效果很明显。[①]

$Y = A^T [[GgG^T] \odot [B^T dB]] A$

以上推广到二维数据，同样可以实现卷积计算的加速。

4.2 硬件层优化

硬件层优化是通过处理器的物理计算单元、指令集、内存资源访问、缓存机制等实现算法的加速。

4.2.1 SIMD 指令向量化

计算机体系结构按照指令流（Instruction）和数据流（Data）两个维度可以分为四种类型（Flynn's Taxonomy，Flynn 分类法）：单指令单数据（Single Instruction Single Data，SISD）、单指令多数据（Single Instruction Multiple Data，SIMD）、多指令单数据（Multiple Instruction Single Data，MISD）、多指令流多数据流（Multiple Instruction Stream Multiple Data，MIMD）。

其中针对 SIMD 类型，计算机可以通过 SIMD 指令集优化程序。SIMD 指令可以实现一条指令操作多个数据，也就是在处理器、寄存器里实现数据的并行操作，它适合数据密集型运算。这样的指令集有 X86 的 MMX［MultiMedia eXtensions，多媒体扩展，如图 4-10（a）所示］、SSE［Streaming SIMD Extensions，流式单指令多数据扩展，如图 4-10（b）所示］、AVX［Advanced Vector Extensions，高级矢量扩展，如图 4-10（c）

---

① 更多可以参考［1］Lavin A，Gray S. Fast Algorithms for Convolutional Neural Networks. IEEE，2015.

［2］Shi F，Li H，Gao Y，et al. Sparse Winograd Convolutional neural networks on small-scale systolic arrays. 2018.

Instruction Set
- ☑ MMX
- ☐ SSE
- ☐ SSE2
- ☐ SSE3
- ☐ SSSE3
- ☐ SSE4.1
- ☐ SSE4.2
- ☐ AVX
- ☐ AVX2
- ☐ FMA
- ☐ AVX_VNNI
- ☐ AVX-512
- ☐ KNC
- ☐ AMX
- ☐ SVML
- ☐ Other

Categories
- ☐ Application-Targeted
- ☑ Arithmetic
- ☐ Bit Manipulation

🔍 Search Intel Intrinsics

```
__m64 _mm_add_pi16 (__m64 a, __m64 b)
__m64 _mm_add_pi32 (__m64 a, __m64 b)
__m64 _mm_add_pi8 (__m64 a, __m64 b)
__m64 _mm_adds_pi16 (__m64 a, __m64 b)
__m64 _mm_adds_pi8 (__m64 a, __m64 b)
__m64 _mm_adds_pu16 (__m64 a, __m64 b)
__m64 _mm_adds_pu8 (__m64 a, __m64 b)
__m64 _mm_madd_pi16 (__m64 a, __m64 b)
__m64 _mm_mulhi_pi16 (__m64 a, __m64 b)
__m64 _mm_mullo_pi16 (__m64 a, __m64 b)
__m64 _m_paddb (__m64 a, __m64 b)
__m64 _m_paddd (__m64 a, __m64 b)
__m64 _m_paddsb (__m64 a, __m64 b)
__m64 _m_paddsw (__m64 a, __m64 b)
__m64 _m_paddusb (__m64 a, __m64 b)
__m64 _m_paddusw (__m64 a, __m64 b)
__m64 _m_paddw (__m64 a, __m64 b)
__m64 _m_pmaddwd (__m64 a, __m64 b)
```

（a）MMX 指令

Instruction Set
- ☐ MMX
- ☑ SSE
- ☐ SSE2
- ☐ SSE3
- ☐ SSSE3
- ☐ SSE4.1
- ☐ SSE4.2
- ☐ AVX
- ☐ AVX2
- ☐ FMA
- ☐ AVX_VNNI
- ☐ AVX-512
- ☐ KNC
- ☐ AMX
- ☐ SVML
- ☐ Other

Categories
- ☐ Application-Targeted
- ☑ Arithmetic
- ☐ Bit Manipulation

🔍 Search Intel Intrinsics

```
__m128 _mm_add_ps (__m128 a, __m128 b)
__m128 _mm_add_ss (__m128 a, __m128 b)
__m128 _mm_div_ps (__m128 a, __m128 b)
__m128 _mm_div_ss (__m128 a, __m128 b)
__m128 _mm_mul_ps (__m128 a, __m128 b)
__m128 _mm_mul_ss (__m128 a, __m128 b)
__m64  _mm_mulhi_pu16 (__m64 a, __m64 b)
__m64  _m_pmulhuw (__m64 a, __m64 b)
__m64  _m_psadbw (__m64 a, __m64 b)
__m64  _mm_sad_pu8 (__m64 a, __m64 b)
__m128 _mm_sub_ps (__m128 a, __m128 b)
__m128 _mm_sub_ss (__m128 a, __m128 b)
```

（b）SSE 指令

（c）AVX 指令

图 4-10　指令集[①]

所示]、ARM 的 Neon [如图 4-11 和图 4-12 所示，关于 Neon 指令类型、使用方法、编程和实例在第 8 章中会进一步讨论]、Armv8 AArch64 架构下一代 SIMD 指令集——SVE 指令集（Scalable Vector Extension）、MIPS 架构的 X-Burst、Qualcomm 的 HVX 等等。

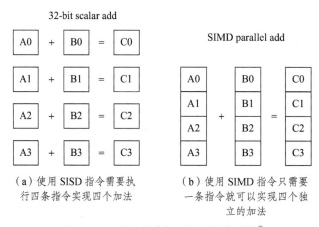

（a）使用 SISD 指令需要执行四条指令实现四个加法
（b）使用 SIMD 指令只需要一条指令就可以实现四个独立的加法

图 4-11　SISD 指令与 SIMD 指令对比[②]

---

① 图来源：Intel® Intrinsics Guide，https://www.intel.com/content/www/us/en/docs/intrinsics-guide/index.html
② 图来源：https://developer.arm.com/documentation/den0013/d/Introducing-NEON/SIMD

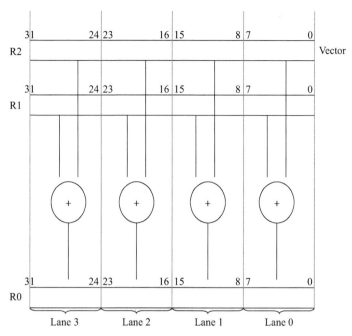

图 4-12　四个 8 位整型数据相加（通过一条 SIMD 指令 UADD8 R0, R1, R2 实现）[1]

4.2.2　多核 CPU 中 OpenMP 编程

OpenMP 是一种实现程序多线程并行运行的方案，其中这些线程共享内存，很适合多核 CPU 处理器架构中的程序并行运行。OpenMP 支持的编程语言有 C/C++、Fortran。基于 OpenMP 实现算法优化具有很好的收益，比如用于图像二值化计算的优化，以下提供完整的测试代码（图 4-13 为测试图片）。

```
//#include <arm_neon.h>
#include <stdio.h>

//#include<iostream>
#include" omp.h"
//#include <arm_sve.h>
#include" opencv2/opencv.hpp"
#include" opencv2/core.hpp"
using namespace cv;
using namespace std;

int thr = 100;
```

---

[1]　图来源：https://developer.arm.com/documentation/den0013/d/Introducing-NEON/SIMD

```cpp
// NO OpenMP
int threshold ( const Mat& src )
{
  int H = src.rows;
  int W = src.cols;
  int sum = 0;
  for ( int h = 0; h < H; ++h )
    for ( int w = 0; w < W; ++w )
      if ( src.at < uchar > ( h, w ) > thr )
        sum += 1;
  return sum;
}

// OpenMP
int threshold_omp ( const Mat& src )
{
  int H = src.rows;
  int W = src.cols;
  int sum = 0;
  #pragma omp parallel for reduction ( +:sum ) num_threads ( 6 )
  for ( int h = 0; h < H; ++h )
    for ( int w = 0; w < W; ++w ) {
      if ( src.at < uchar > ( h, w ) > thr )
        sum += 1;
        //int id = omp_get_thread_num ( );
        //int rank = omp_get_num_threads ( );
        //cout <<" thread" << id <<" of" << rank << endl;
    }
  return sum;
}

#define NUM 80
int main ( ) {
  Mat gray_, out_1, out_2;
  Mat src = imread ( " test.jpg" );
```

```cpp
int sum = 0;
cout <<" W:" << src.cols <<" H:" << src.rows << endl;
cout <<" 计算耗时 NO OpenMP | OpenMP" << endl;
cvtColor（src, gray_, COLOR_BGR2GRAY）;
double t_1 = static_cast<double>（getTickCount（））;
for（int i =0; i<NUM; i++）
  sum = threshold（gray_）;
t_1 =((double) getTickCount（）- t_1) / getTickFrequency（）;
cout << sum <<" \n" << endl;
sum = 0;
cvtColor（src, gray_, COLOR_BGR2GRAY）;
double start = omp_get_wtime（）;

for（int i = 0; i < NUM; i++）{
  //int my_id = omp_get_thread_num（）;
  //int my_rank = omp_get_num_threads（）;
  //std::cout <<" Hello from thread" << my_id <<" of" << my_rank << std::endl;
  sum = threshold_omp（gray_）;
}
double end = omp_get_wtime（）;
double t_2 = end −start;
cout << sum << endl;
cout << t_1/NUM <<" " << t_2/NUM <<" s" << endl;
//cout <<" \n" << endl;

//imwrite（" src.jpg", src）;
return 0;
}
```

代码编译后测试命令如下：

CPU 核数

（base）user@user:/testneon$ cat /proc/cpuinfo| grep" cpu cores" | uniq

cpu cores   : 6

（base）user@user:/testneon$ pkg-config opencv4 --cflags

-I/usr/local/include/opencv4

（base）user@user:/testneon$ pkg-config opencv4 --libs

-L/usr/local/lib -lopencv_gapi -lopencv_highgui -lopencv_ml -lopencv_objdetect -lopencv_photo -lopencv_stitching -lopencv_video -lopencv_calib3d -lopencv_features2d -lopencv_dnn -lopencv_flann -lopencv_videoio -lopencv_imgcodecs -lopencv_imgproc -lopencv_core

（base）user@user:/testneon$ g++ \`pkg-config opencv4 --cflags\` obj_intel.c -o obj_intel \`pkg-config opencv4 --libs\` -std=c++11 -fopenmp -lgomp

（base）user@user:/testneon$ ./obj_intel

W: 3968 H: 2976

计算耗时 NO OpenMP | OpenMP

4527870

4527870

0.0504567 0.0104724 s

图 4-13 测试图片（照片拍摄于古老又迷人的凤凰古城。让我们一起去感受沈从文先生笔下《边城》中的凤凰古城"倚山而筑，环以石墙，濒临沱江，群山环抱，河溪萦回，关隘雄奇"）

通过以上的测试，我们可以得到 OpenMP 在这里具有至少 5 倍的加速效果。

下面我们在硬件芯片 V853 Arm Cortex-A7 CPU core @ 1 GHz 中测试这段代码。首先在代码中添加 #include <arm_neon.h> 头文件以及配置改为：#pragma omp parallel for reduction（+:sum）num_threads（2）。

用命令查看 V853 中 ARMv7 核的硬件资源配置：

root@TinaLinux:/# cat /proc/cpuinfo

processor: 0

model name: ARMv7 Processor rev 5（v7l）

BogoMIPS: 48.00

Features: half thumb fastmult vfp edsp neon vfpv3 tls vfpv4 idiva idivt vfpd32 lpae

CPU implementer: 0x41

CPU architecture: 7

CPU variant: 0x0

CPU part: 0xc07

CPU revision: 5

Hardware: sun8iw21

Revision: 0000

Serial: 0000000000000000

root@TinaLinux:/#

再通过以下的 Shell 命令实现自动化编译和测试。

#!/bin/bash

export V853_DIR=/……/V853

export OBJ=obj

echo″ $V853_DIR″

export TOOLCHAIN_BIN_DIR=/tina-v853-open/prebuilt/rootfsbuilt/arm/toolchain-sunxi-musl-gcc-830/toolchain/bin/

export STAGING_DIR=${V853_DIR}/tina-v853-open/prebuilt/rootfsbuilt/arm/toolchain-sunxi-musl-gcc-830/toolchain/arm-openwrt-linux-muslgnueabi

rm ${OBJ}

printf″ rm ${OBJ} \n\n″

#${V853_DIR}${TOOLCHAIN_BIN_DIR}/arm-openwrt-linux-gcc -g -o ${OBJ} ${OBJ}.c

${V853_DIR}${TOOLCHAIN_BIN_DIR}/arm-openwrt-linux-g++ ${OBJ}.c -I ./opencv4 -L ./lib -lopencv_core -lopencv_imgcodecs -lopencv_imgproc -fopenmp -o ${OBJ}

file ${OBJ}

readelf -a ${OBJ} | grep NEEDED

printf″ \n\n″

printf″ ls \n″

ls

printf″ adb push ${OBJ} /tmp \n″

adb push ./lib /tmp/libcv

adb push test.jpg /tmp

adb push ${OBJ} /tmp

#export LD_LIBRARY_PATH=./libcv:$LD_LIBRARY_PATH

最后我们得到的测试结果如下：

root@TinaLinux:/tmp# export LD_LIBRARY_PATH=./libcv:$LD_LIBRARY_PATH

root@TinaLinux:/tmp# ./obj

W: 3968 H: 2976

计算耗时 NO OpenMP | OpenMP

4527870

4527870

0.858552 0.915989 s

root@TinaLinux:/tmp# exit

测试发现性能没有提升，反而有点下降。

我们再在编译中配置 -O3 后，得到的测试结果如下：

root@TinaLinux:/tmp# ./obj

W: 3968 H: 2976

计算耗时 NO OpenMP | OpenMP

4527870

4527870

0.0412356 0.0762503 s

OpenMP 适用于多核的处理器架构，对于单核往往不能起到加速的效果。

4.2.3　GPU 并行计算

并行计算是计算机同时进行多个计算任务的模式，与其对应的是串行计算。并行计算按照颗粒度可分为处理器位级并行、指令集并行和任务级并行等。处理器位级并行是指处理器的位宽大小，可分为 8 位、16 位、16 位、32 位、64 位等。简单举例说明一下：如果要实现 32 位的数据加法，对于 8 位处理器，需要多次 8 位计算才可以，而 32 位处理器只需要计算一步。所以位宽高的处理器一般比位宽低的处理器快。指令集并行指前文讨论的 SIMD。任务级并行可以理解为将一个任务分解为多个可以并行执行的相同或不同的微型任务。目前用于并行计算的框架有 CPU-OpenMP、GPU-CUDA、OpenCL、Vulkan 等。

GPU 并行计算是基于通用 GPU 核并行执行由大任务分解的多个子任务的方法。

GPU是计算机显卡的核心芯片。刚开始显卡的功能主要是显示渲染，现在显卡芯片不仅可以实现显示渲染，更能实现通用并行加速计算。

在手机等智能终端设备中，目前的主流GPU主要有PowerVr、Adreno、Mali，其中Adreno GPU是Qualcomm Snapdragon处理器的GPU核，Mali是ARM的高端GPU。

在嵌入式边缘设备中GPU方案有NVIDIA Jetson（如图4-14所示）和NVIDIA DRIVE。NVIDIA Jetson是一个完整的系统模组，它包括CPU、GPU、内存、电源管理和高速接口等，有Jetson TX2、Jetson Xavier、Jetson Nano等系列。NVIDIA DRIVE硬件有NVIDIA DRIVE Hyperion、NVIDIA DRIVE Orin、NVIDIA DRIVE AGX Pegasus、NVIDIA DRIVE AGX Xavier、NVIDIA DRIVE Atlan。其中NVIDIA DRIVE Orin™ SoC（系统级芯片）可提供每秒254 TOPS（万亿次运算），是智能车辆的中央计算机。

图4-14　NVIDIA Jetson[①]

GPU并行计算是通过可编程方式充分利用GPU内部并行单元资源去提高算法处理性能。NVIDIA GPU的架构包括主机端（Host）和设备端（Device），如图4-15所示。

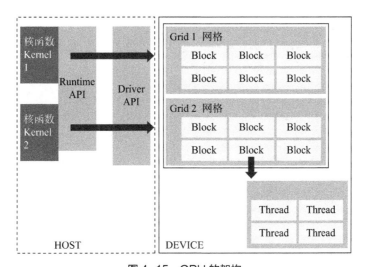

图4-15　GPU的架构

---

① 图来源：https://www.nvidia.cn

CUDA（统一计算架构）是 NVIDIA 推出的基于 GPU 的通用并行计算架构，包括 CUDA 指令集架构和 GPU 内部的并行计算引擎。CUDA 作为一种异构计算平台，通过 CUDA 编程可以最大化和自主化使用 GPU 的并行计算能力来优化我们的算法。CUDA 编程包括两部分代码：主机端（host 代码，在 CPU 中运行，负责分配内存和复制数据）和设备端（device 代码，在 GPU 中运行的并行代码，称为 Kernel 函数）。

在设备端运行的 Kernel 函数为并行处理函数，也叫核函数。NVCC 是编译 CUDA 程序的编译器。以下提供一个完整的 CUDA kernel 测试程序示例，功能是 CUDA kernel 优化图片二值化处理和 Sobel 计算。

```cpp
#include <iostream>
#include <string>
#include <cassert>

#include <cuda.h>
//#include "cuda_runtime.h"
//#include "device_launch_parameters.h"
//#include <cuda_runtime_api.h>
//#include <device_functions.h>

//OpenCV 实现图片的加载和像素处理
#include <opencv2/opencv.hpp>
#include <iostream>
using namespace std;
using namespace cv;

#define checkCudaErrors( val ) check ( ( val ), #val, __FILE__, __LINE__ )

cv::Mat img_rgba;
cv::Mat img_grey;

//GPU memory
uchar4 *d_rgba_img_;
uchar *d_grey_img_;

template<typename T>
void check ( T err, const char* const func, const char* const file, const int line )
{
    if( err != cudaSuccess )
```

第4章 算子和图优化

```
        {
            std::cerr <<" CUDA error at:" << file <<" :" << line << std::endl;
            std::cerr << cudaGetErrorString（err）<<" " << func << std::endl;
            exit（1）;
        }
    }
}

// 图片预处理
void data_preparation（uchar4 **input_img, unsigned char **grey_img, uchar4 **d_rgba_img, unsigned char **d_grey_img, cv::Mat img）
{
    checkCudaErrors（cudaFree（0））;
    // 释放指针指向的显存；如果输入参数为 0，不进行释放操作，后面程序不会再次初始化
    int h = img.rows;
    int w = img.cols;

    //BGR 格式转为 RGBA 格式
    cv::cvtColor（img, img_rgba, COLOR_BGR2RGBA）;

    // 生成和原图一样大小的 img_grey
    img_grey.create（img.rows, img.cols, CV_8UC1）;

    // 判断图像是否连续存放
    if（!img_rgba.isContinuous（）|| !img_grey.isContinuous（））{
        std::cerr <<" Images aren't continuous!! Exiting." << std::endl;
        exit（1）;
    }

    //input_img 指向 img_rgba
    *input_img =（uchar4 *）img_rgba.ptr<unsigned char>（0）;
    //grey_img 指向 img_grey
    *grey_img = img_grey.ptr<unsigned char>（0）;

    // 分配 GPU memory
    const size_t img_size = h*w;
checkCudaErrors（cudaMalloc（d_rgba_img, sizeof（uchar4）*img_size））;
// cudaMalloc 函数分配 GPU 内存
```

```cpp
    checkCudaErrors（cudaMalloc（d_grey_img, sizeof（unsigned char）*img_size））;
    //cudaMemset 函数将 GPU 内存空间 d_grey_img 置 0
     checkCudaErrors（cudaMemset（*d_grey_img, 0, img_size * sizeof（unsigned char））））;

    // cudaMemcpy 函数将 input_img 数据复制给 GPU 内存空间 d_rgba_img
    checkCudaErrors（cudaMemcpy（*d_rgba_img, *input_img, sizeof（uchar4）*img_size, cudaMemcpyHostToDevice））;

    d_rgba_img_ = *d_rgba_img;
    d_grey_img_ = *d_grey_img;
}

//rgba 转换为灰度图核函数
__global__ void rgba2grey_cuda（const uchar4* const rgba_img, unsigned char* const grey_img, int h, int w）
{
int threadId = blockIdx.x * blockDim.x * blockDim.y +
threadIdx.y * blockDim.x +
threadIdx.x;
/*
CUDA 软件架构由网格 Grid、线程块 Block 和线程 Thread 组成
threadIdx：线程索引；
blockIdx：线程块索引，一个线程块包含多个线程；
blockDim：线程块大小；
gridDim：网格大小，一个网格包含多个线程块；
blockDim.x（y、z）代表线程块在 x、y、z 方向上的线程数；
gridDim.x（y、z）代表网格里沿 x（y、z）方向的线程块数
*/
    if（ threadId < h * w）
    {
        const unsigned char r = rgba_img［threadId］.x;
        const unsigned char g = rgba_img［threadId］.y;
        const unsigned char b = rgba_img［threadId］.z;
        grey_img［threadId］= .299f*r + .587f*g + .114f*b;
```

            }
        }

    //Sobel算子边缘检测核函数
        __global__ void sobel_cuda（unsigned char *data_in, unsigned char *data_out, int h, int w）
        {
            int x_index = threadIdx.x + blockIdx.x * blockDim.x;
            int y_index = threadIdx.y + blockIdx.y * blockDim.y;
            int index = y_index * w + x_index;
            int Gx = 0;
            int Gy = 0;

            if（x_index > 0 && x_index < w - 1 && y_index > 0 && y_index < h - 1）
            {
                Gx = data_in［(y_index - 1) * w + x_index + 1］+
                    2 * data_in［y_index * w + x_index + 1］+
                    data_in［(y_index + 1) * w + x_index + 1］-
                    (data_in［(y_index - 1) * w + x_index - 1］+
                    2 * data_in［y_index * w + x_index - 1］+
                    data_in［(y_index + 1) * w + x_index - 1］);

                Gy = data_in［(y_index - 1) * w + x_index - 1］+
                    2 * data_in［(y_index - 1) * w + x_index］+
                    data_in［(y_index - 1) * w + x_index + 1］-
                    (data_in［(y_index + 1) * w + x_index - 1］+
                    2 * data_in［(y_index + 1) * w + x_index］+
                    data_in［(y_index + 1) * w + x_index + 1］);

                data_out［index］=（abs（Gx）+ abs（Gy））/ 2;
            }
        }

    //Sobel算子边缘检测CPU函数
        void sobel（Mat src, Mat dst, int h, int w）
        {

```cpp
    int Gx = 0;
    int Gy = 0;
    for ( int i = 1; i < h - 1; i++ )
    {
        uchar *data_up = src.ptr <uchar> ( i - 1 );
        uchar *data = src.ptr <uchar> ( i );
        uchar *data_down = src.ptr <uchar> ( i + 1 );

        uchar *out = dst.ptr <uchar> ( i );

        for ( int j = 1; j < w - 1; j++ )
        {
            Gx = ( data_up [ j + 1 ] + 2 * data [ j + 1 ] + data_down [ j + 1 ] ) -
            ( data_up [ j - 1 ] + 2 * data [ j - 1 ] + data_down [ j - 1 ] );

            Gy = ( data_up [ j - 1 ] + 2 * data_up [ j ] + data_up [ j + 1 ] ) -
            ( data_down [ j - 1 ] + 2 * data_down [ j ] + data_down [ j + 1 ] );

            out [ j ] = ( abs ( Gx ) + abs ( Gy ) ) / 2;
        }
    }
}

int main ( int argc, char* argv [ ] )
{
    // 输入地址
    std::string in_img_path = " in_.jpg" ;
    // 输出地址
    std::string out_img_path = " out_.jpg" ;

    //host 指针、device 指针
    uchar4 *h_rgba_img, *d_rgba_img;
    unsigned char *h_grey_img, *d_grey_img;

    // 读取图片
    cv::Mat img = cv::imread ( in_img_path.c_str ( ),
    IMREAD_COLOR ); //CV_LOAD_IMAGE_COLOR
```

```
if( img.empty（ ） ){
  std::cerr <<" Couldn't open file:" << std::endl;
  exit（1）;
}
int h = img.rows;
int w = img.cols;
// 数据赋值给 h_rgbaImage，复制给 d_greyImage
data_preparation（&h_rgba_img, &h_grey_img, &d_rgba_img, &d_grey_img, img）;

// 并行处理 Kernel
int thread = 32;
int grid =（h*w + thread − 1）/（thread*thread）;
const dim3 blockSize（thread, thread）;
const dim3 gridSize（grid）;

// 启动在 GPU 设备上计算的线程，<<<>>> 运算符内是核函数的执行参数
// <<<gridSize, blockSize>>> 代表 gridSize 个线程块，每个线程块包含 blockSize 个并行线程
rgba2grey_cuda <<<gridSize, blockSize >>>（d_rgba_img, d_grey_img, h, w）;

// 阻塞
cudaDeviceSynchronize（）;

//GPU 结果复制给 CPU
size_t imgsize = h*w;
  checkCudaErrors（cudaMemcpy（h_grey_img, d_grey_img, sizeof（unsigned char）
*imgsize, cudaMemcpyDeviceToHost））;

// 写入图片
cv::Mat output（h, w, CV_8UC1, h_grey_img）;
cv::imwrite（out_img_path.c_str（）, output）;

// 显存释放
cudaFree（d_rgba_img_）;
cudaFree（d_grey_img_）;

//========Sobel 算子处理 ==========================
Mat gray_img = output; //imread（"1.jpg", 0）;
```

```cpp
Mat gauss_img;
// 高斯滤波
GaussianBlur（gray_img, gauss_img, Size（3, 3）, 0, 0, BORDER_DEFAULT）;

//Sobel算子CPU实现
Mat dst（h, w, CV_8UC1, Scalar（0））;
sobel（gauss_img, dst, h, w）;

//CUDA实现后的传回的图像
Mat dst_img（h, w, CV_8UC1, Scalar（0））;

// 创建GPU内存
unsigned char *d_in;
unsigned char *d_out;

cudaMalloc（（void**）&d_in, h * w * sizeof（unsigned char））;
cudaMalloc（（void**）&d_out, h * w * sizeof（unsigned char））;

// 将高斯滤波后的图像从CPU传入GPU
    cudaMemcpy（d_in, gauss_img.data, h * w * sizeof（unsigned char）,
cudaMemcpyHostToDevice）;

    dim3 threadsPerBlock（32, 32）;
    dim3 blocksPerGrid（（w + threadsPerBlock.x - 1）/ threadsPerBlock.x,（h +
threadsPerBlock.y - 1）/ threadsPerBlock.y）;
    /*
    */
    // 调用核函数
    sobel_cuda <<<blocksPerGrid, threadsPerBlock >>> （d_in, d_out, h, w）;

    // 将图像传回GPU
    cudaMemcpy（dst_img.data, d_out, h * w * sizeof（unsigned char）,
cudaMemcpyDeviceToHost）;

    //cv::Mat output（h, w, CV_8UC1, h_greyimg）;
    std::string sobel_img ="sobel_img.jpg";
    cv::imwrite（sobel_img.c_str（）, dst_img）;
```

```
    // 释放 GPU 内存
    cudaFree ( d_in );
    cudaFree ( d_out );
}
```

以上程序在实现图片的二值化处理和 Sobel 的 GPU 计算中为图片每个像素启动一个线程来实现各像素的并行计算。CUDA 的操作主要步骤如下：

（1）使用函数 cudaMalloc 实现在 GPU 上申请内存：

float* inD, * outD;

cudaMalloc ((void**) &inD, sizeof (float)* N);

其中 inD 参数传递的是 inD 指针的地址，也就是指针的指针或二重指针。

（2）使用函数 cudaMemcpy 实现主机（Host）和设备（Device）之间数据的传递：

cudaMemcpy (res_1D, outD, sizeof (float)* N, cudaMemcpyDeviceToHost);

该函数是同步执行函数，在执行中会锁死并占有 CPU 进程的控制权，直到完成数据的传递；其中参数 cudaMemcpyHostToDevice 代表 Host 到 Device，cudaMemcpyDeviceToHost 代表 Device 到 Host。

（3）在 GPU 上启动 kernel，以上测试代码中已经实现了 kernel 函数。关于这个 CUDA C 的测试程序编译方法如下：

nvcc -o testcuda.out testcuda.cu -I/usr/local/include/opencv4 -lopencv_highgui -lopencv_imgcodecs -lopencv_imgproc -lopencv_core

测试代码中的函数 cudaDeviceSynchronize 的作用是阻塞当前程序的执行，直到所有线程执行完成 kernel 函数。

（4）使用函数 cudaMemcpy 把数据从 GPU 取回。

（5）使用函数释放 GPU 内存。

针对 ARM 的 Mali GPU 和 Adreno，我们可以使用 OpenGL、OpenCL、Vulkan 实现操作底层 GPU 资源来加速计算。

CUDA C 语言编程是算法在 GPU 上实现并行加速的重要方法，鉴于本书篇幅的限制，不再对 CUDA C 编程做进一步的讨论，读者朋友如果想深入理解和掌握 CUDA C 编程，可以参考学习相关的 CUDA 编程书籍。

接下来简单介绍一下基于 OpenGL、OpenCL 或 Vulkan 实现在移动端平台中 GPU 并行加速。

OpenGL（Open Graphics Library，开放图形库）是用于渲染 2D、3D 矢量图形的跨语言和跨平台的应用程序编程接口（API）。[1]

---

[1] https://www.opengl.org/

OpenCL（Open Computing Language，开放运算语言）是异构平台通用并行编程的开放式标准，如图 4-16 所示。

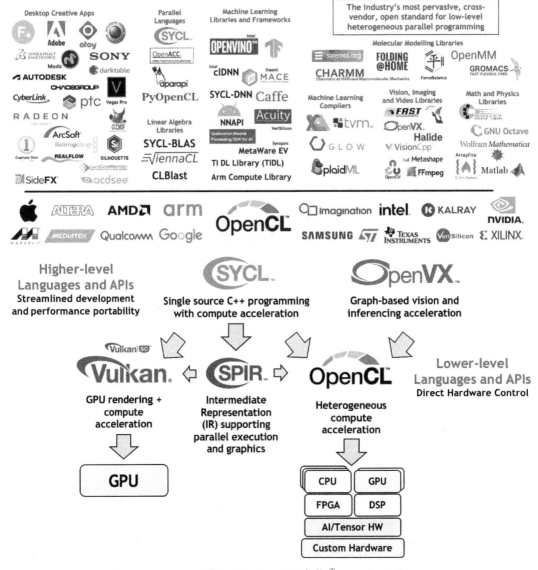

图 4-16　OpenCL 架构[①]

Vulkan 是一套适用于高性能 3D 图形的低开销、跨平台 API。与 OpenGL ES（GLES）一样，Vulkan 提供在应用中创建高品质实时图形的工具。Vulkan 实现直接控制和访问底层 GPU，进而降低 CPU 开销，提高控制底层 GPU 硬件的效率和性能。[②]

---

① 图来源：https://www.khronos.org/opencl/
② 摘自：https://source.android.google.cn/devices/graphics/implement-vulkan，http://vulkan.gpuinfo.org/listdevices.php?platform=linux

Vulkan 机制包含全局工作组、本地工作组、执行单元,其中每一个执行单元执行一次计算着色器,如图 4-17 所示。

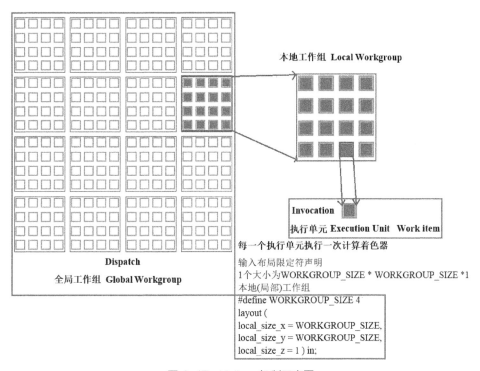

图 4-17　Vulkan 机制示意图

4.2.4　Cache 优化

以上的方法偏向于计算耗时的优化,本节讨论的方法针对访存时延的优化——Cache 优化。

Cache 是 CPU 和内存(一般为 DRAM,Dynamic Random Access Memory)之间容量小但速度很高的高速缓冲存储器,其速度为 ns 级别,接近 CPU 寄存器。Cache 一般用静态存储器 SRAM(Static Random Access Memory)实现。

Cache 分为 L1 Cache 和 L2 Cache。L1 Cache 为哈佛(Harvard)结构,它的指令和数据分开存储,分为 Instruction Cache(I-Cache)和 Data Cache(D-Cache),CPU 能并行访问 I-Cache 和 D-Cache。L2 Cache 为普林斯顿(Princeton)结构,它的指令和数据存储不分开。

多核处理器中每个 CPU 核有一个 L1 Cache,多个 CPU 核共享一个 L2 Cache。

Cache 使用固定大小的块(Block)作为基本单元来存储数据,也叫 Cache Line/Block。

CPU 在计算中一般先在 Cache 中查找指令和数据,再访问内存。如果数据在 Cache

中，则 Cache 命中（Cache hit），反之为缺失（Cache miss）。[①] 因为 Cache 的容量一般很小，当一个数据块很大的时候，Cache 便不能缓存它，所以 Cache 缺失。

在矩阵乘法（Matrix Multiplication）中，将大矩阵分解成小矩阵再参加计算，这些小块矩阵便很容易缓存在 Cache 里，便于后面的继续使用，也就是提高了 Cache 的命中率。以下提供一个完整的矩阵乘法测试代码，以便于读者朋友理解如何分解大矩阵。

```c
#include <string.h>
#include <stdio.h>
#include <stdlib.h>
#include <time.h>
#include <assert.h>

// z=x*y, x:n*m, y:m*s
void mul_standard ( float* z, float* x, float* y, int n, int m, int s ) {
    int i, j, k;
    for ( i = 0; i < n; i++ )
    for ( j = 0; j < s; j++ )
    for ( k = 0; k < m; k++ )
    z [ j + i*s ] += x [ k + i*m ] * y [ j + k*s ];
}

#define BLOCK_SIZE 16
void mul_block ( float* z, float* x, float* y, int n, int m, int s ) {
    int i, j, k, i_, j_, k_;
    for ( i = 0; i < n; i += BLOCK_SIZE )
    for ( j = 0; j < s; j += BLOCK_SIZE )
    for ( k = 0; k < m; k += BLOCK_SIZE )
     for ( i_ = i; i_ < i + BLOCK_SIZE && i_ < n; i_++ )
      for ( j_ = j; j_ < j + BLOCK_SIZE && j_ < s; j_++ )
       for ( k_ = k; k_ < k + BLOCK_SIZE && k_ < m; k_++ )
       z [ j_ + i_*s ] += x [ k_ + i_*m ] * y [ j_ + k_*s ];
}

#define N 640
```

[①] 更多资料可以参考 Cache Blocking Techniques，https://www.intel.cn/content/www/cn/zh/developer/articles/technical/cache-blocking-techniques.html

```c
#define M 320
#define S 640

void main ( ) {
    float* x = ( float* ) malloc ( N*M*sizeof ( float ) );
    float* y = ( float* ) malloc ( M*S*sizeof ( float ) );

    int i, j;
    for ( i=0; i<N; i++ )
      for ( j=0; j<M; j++ )
        x [ i*M+j ] = 3.14159;

    for ( i=0; i<M; i++ )
      for ( j=0; j<S; j++ )
        y [ i*S+j ] = 1.414;

    float* z1 = ( float* ) calloc ( N*S, sizeof ( float ) );

    clock_t start, end;
    start = clock ( );
    mul_standard ( ( float* ) z1, ( float* ) x, ( float* ) y, N, M, S );
    end = clock ( ) - start;
printf ( " mul_standard cost time: %ld ms\n" , end*1000/CLOCKS_PER_SEC );

    float* z2 = ( float* ) calloc ( N*S, sizeof ( float ) );
    //memset ( z, 0, N*S*sizeof ( float ) );
    start = clock ( );
    mul_block ( ( float* ) z2, ( float* ) x, ( float* ) y, N, M, S );
    end = clock ( ) - start;
    printf ( " mul_block cost time: %ld ms\n" , end*1000/CLOCKS_PER_SEC );

    int flag = 0;
    for ( i=0; i<N; i++ )
      for ( j=0; j<S; j++ )
        if ( z1 [ i*S+j ] != z2 [ i*S+j ] ) {
          flag = -1;
          //printf ( " z1 z2 : %f, %f\n" , z1 [ i*S+j ], z2 [ i*S+j ] );
```

```
      } //else
      //printf（"z1 z2 : %f, %f\n", z1［i*S+j］, z2［i*S+j］）;

    if（flag == -1）
      printf（"test failed\n"）;
    else
      printf（"test success\n"）;

    free（x）;
    free（y）;
    free（z1）;
    free（z2）;
    return;
}
```

以上代码编译和测试得到的结果如下：

```
$ gcc block_mul.c -o mul_test.out -march=native -Ofast -funroll-loops && ./mul_test.out
mul_standard cost time: 187 ms
mul_block cost time: 126 ms
```

矩阵 x 和 y 乘中，如果矩阵的维度很大，那么 x、y 会大于 Cache 的容量，导致 CPU 计算中 Cache 命中率很低，加大内存的访问量，导致访存时间开销很大。通过矩阵分块（Blocking）将大矩阵分解成小的子矩阵，可以提高访存指令从 Cache 中直接读取到数据的命中率，从而减小平均访存指令延迟时间。

我们可以进一步通过 Perf（Performance Event）工具去分析 Cache 优化。Perf 工具是 Linux kernel 自带的程序性能分析工具（profiling 工具），它能够对程序从进程、线程到函数级再到指令进行跟踪和分析。对于 Ubuntu 系统，如果没有安装 Perf，可以通过命令 sudo apt-get install linux-tools-common linux-tools-generic 自行安装。Perf 功能非常强大，可以收集 CPU 运行中如 CPU 利用率、内存使用情况、Cache 缓存命中率、指令执行次数等与性能相关的多个重要数据，并实现函数调用图、火焰图等可视化显示功能，方便我们定位到程序。Perf 用于分析 Cache 缓存命中率的具体方法如下：

```
$ g++ test_cache1.c -o test_cache1
$ sudo perf stat -d -d ./test_cache1
test_1 cost=0.185033

 Performance counter stats for'./test_cache1'：
```

```
           194.92      msec task-clock           # 0.996 CPUs utilized
                9      context-switches          # 0.046 K/sec
                4      cpu-migrations            # 0.021 K/sec
           51,329      page-faults               # 0.263 M/sec
      671,440,104      cycles                    # 3.445 GHz
(26.35%)
    1,229,844,034      instructions              # 1.83 insn per cycle
(34.54%)
      129,320,600      branches                  # 663.454 M/sec
(36.59%)
          323,895      branch-misses             # 0.25% of all branches
(38.66%)
      354,793,393      L1-dcache-loads           # 1820.199 M/sec
(40.70%)
        3,802,075      L1-dcache-load-misses     # 1.07% of all L1-dcache hits
(41.01%)
           48,819      LLC-loads                 # 0.250 M/sec
(32.81%)
           28,589      LLC-load-misses           # 58.56% of all LL-cache hits
(32.81%)
    <not supported>    L1-icache-loads
           67,365      L1-icache-load-misses
(32.82%)
      369,262,227      dTLB-loads                # 1894.429 M/sec
(32.64%)
          109,820      dTLB-load-misses          # 0.03% of all dTLB cache hits
(30.60%)
            2,922      iTLB-loads                # 0.015 M/sec
(28.53%)
          101,983      iTLB-load-misses          # 3490.18% of all iTLB cache hits
(26.48%)

      0.195674939 seconds time elapsed

      0.135615000 seconds user
```

0.059830000 seconds sys

$ g++ test_cache2.c -o test_cache2
$ sudo perf stat -d -d ./test_cache2
test_2 cost=0.993135

Performance counter stats for ' ./test_cache2 ':

998.67 msec	task-clock	# 0.999 CPUs utilized
5	context-switches	# 0.005 K/sec
1	cpu-migrations	# 0.001 K/sec
20,608	page-faults	# 0.021 M/sec
3,575,531,583	cycles	# 3.580 GHz
(30.73%)		
1,144,764,761	instructions	# 0.32 insn per cycle
(38.74%)		
117,829,584	branches	# 117.987 M/sec
(39.13%)		
313,003	branch-misses	# 0.27% of all branches
(39.13%)		
340,729,652	L1-dcache-loads	# 341.185 M/sec
(39.12%)		
58,194,680	L1-dcache-load-misses	# 17.08% of all L1-dcache hits
(38.83%)		
1,597,526	LLC-loads	# 1.600 M/sec
(30.43%)		
132,407	LLC-load-misses	# 8.29% of all LL-cache hits
(30.43%)		
\<not supported\>	L1-icache-loads	
116,486	L1-icache-load-misses	
(30.44%)		
336,906,685	dTLB-loads	# 337.356 M/sec
(30.44%)		
65,751	dTLB-load-misses	# 0.02% of all dTLB cache hits
(30.44%)		

（30.44%）	791	iTLB-loads	# 0.792 K/sec
（30.44%）	4，507	iTLB-load-misses	# 569.79% of all iTLB cache hits

```
0.999180862 seconds time elapsed
0.963203000 seconds user
0.035970000 seconds sys
```

4.3 图优化

图优化是对深度学习模型中以子图为颗粒度去优化访存开销和计算量的方法，比如常量折叠（Constant Folding）、算子融合（Operator Fusion）、子图等价替换、子图合并等。

算子融合的一种方法是 BN 折叠（Batch Normalization Folding）。所谓 BN 折叠是指当 Conv 算子后面有一个 BN 算子（批标准化）时，可以将这两个算子融合为一个算子。如图 4-18（a）为 BN+ 卷积的训练图，可以看出在模型训练阶段 BN 中各个参数的形式；图 4-18（b）为 BN+ 卷积的推理图，可以看出 BN 和 Conv2D 做了参数融合；图 4-18（c）为以下代码定义的模型结构，在图中 Conv2D、BatchNormalization 在未融合前还是独立的模块。

```
inputs = tf.keras.layers.Input（shape =（640，320，3））
        x = layers.Conv2D（12，kernel_size=3，strides=1,
        padding='SAME'）(inputs）
x = layers.BatchNormalization（）(x）
outputs = layers.ReLU（）(x）
model = tf.keras.Model（inputs=inputs，outputs=outputs）
model.save（'model_test.h5'）
```

图 4-18（d）为硬件部署的模型结构，在图中 Conv2D 和 BatchNormalization 已经融合为一个模块。本来是两个算子，算子之间需要输入和输出操作，经过 BN 折叠后可以减少内存存储和访问的次数，从而提升部署模型的推理性能。

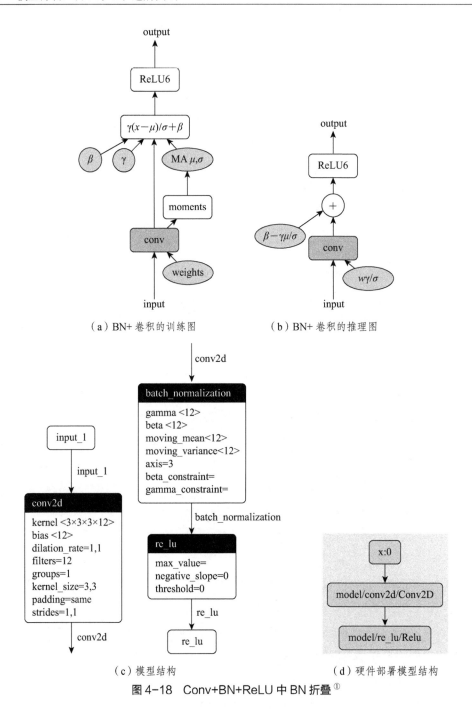

图 4-18 Conv+BN+ReLU 中 BN 折叠[1]

算子融合的另一种方法是，当 Conv 算子后面接一个 Add 或 Mul 算子时，我们可以把这两个融合为一个 Conv 算子，也能实现部署模型性能的提升。

---

[1] 其中图（a）、（b）来源：Jacob B, Kligys S, Chen B, et al.Quantization and Training of Neural Networks for Efficient Integer-Arithmetic-Only Inference. 2017.DOI:10.48550/arXiv.1712.05877.

4.4 AI 编译器

前面讨论的优化方法的共同特点是手工优化，这个对于小型网络往往容易胜任，毕竟算子不是那么多，工程师还是可以 hold 住的。但是当模型网络结构变得更加复杂和巨大的时候，特别是当前的大模型（"大得离谱"），面对这些模型中大量的算子，对于工程师就显得有点力不从心了。AI 编译器（AI Compiler）就是为了实现自动去做算子和图优化，突破手工优化的瓶颈。

讨论 AI 编译器之前，首先讲一下什么是编译器。所谓编译器就是将 C、C++、Java 等高级语言（High-level Language）翻译成机器语言，一般编译的过程如图 4-19 所示，包括预处理、编译、汇编、链接。

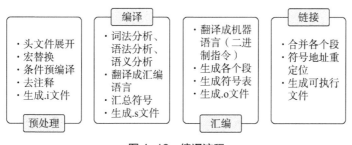

图 4-19 编译流程

AI 编译器就是将 AI 模型计算图（IR）通过编译器的方法实现特定硬件平台的性能优化，包括图优化、算子优化、指令调度、片上存储器复用等，最终得到目标机器码。图 4-20 所示为 AI 编译器流程，它包含前端优化和后端优化。其中前端负责与硬件无关的图优化（High-level IR），而后端负责与硬件紧密相关的图优化（Low-level IR）以及目标代码生成。

图优化中较常见的是通过算子融合来减小访存开销，进而实现计算性能优化。

目前 AI 编译器框架有 Glow、TensorFlow XLA[①]、Tensor Comprehension、MLIR、TVM（Tensor Virtual Machine）、Halide。其中 TVM 是一个开源的端到端的深度学习模型编译框架，TVM 通过图优化及算子优化加速深度学习模型在 CPU、GPU、DSP、NPU、ARM 等目标平台中的推理性能，它的架构如图 4-21 所示。

---

① XLA：https://tensorflow.google.cn/xla/architecture?hl=zh-cn

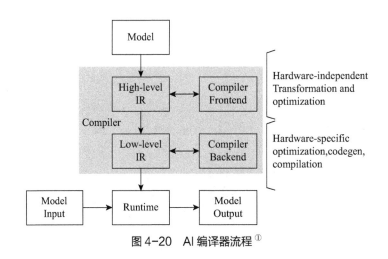

图 4-20　AI 编译器流程[1]

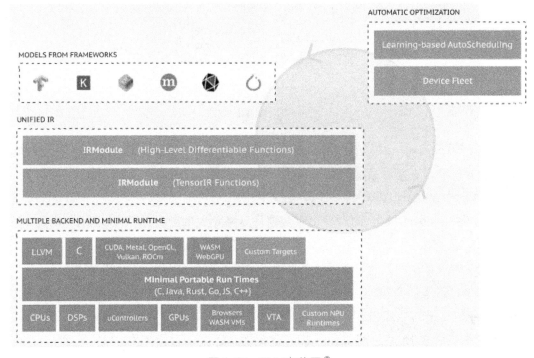

图 4-21　TVM 架构图[2]

---

① 图来源：https://bbs.huaweicloud.com/blogs/351263
② 图来源：https://tvm.apache.org/

第 5 章

模型压缩

本章讨论模型压缩。什么是模型压缩呢？

我们知道模型在端侧的运行会受到存储空间、算力和功耗的限制，所以需要将待部署的模型通过一定的方法实现模型体积变得更小、推理时延更小和低功耗的需求，前提是模型的精度损失满足需求。以上的任务便是模型压缩，其内容涉及轻量化网络模型设计、剪枝、网络架构搜索、低秩分解、知识蒸馏和量化等。特别提醒：本章内容是端侧AI模型部署的必备知识，其中轻量化网络模型设计是入门级内容，讨论得比较基础。通过模型压缩实现模型端侧部署，"让AI能力下沉到端侧，给我们创造更多的想象力"。

正文开始前，请允许笔者先和大家聊一聊"量产经验"这个关键词——这叫开头彩蛋，预热一下氛围。量产经验是开发的产品在市场上得到检验的经验，它是工程师的重要竞争力。对于工程师，做了一大堆的DEMO或预研项目，也抵不过一个量产产品。产品从开发到进入市场，这个过程中遇到的难点或者叫"坑"，小到芯片选型、电路PCB设计、一行关键代码，大到团队建设，都没有想当然的"简单"。乘风破浪的工程师们，一路上攻克技术难点和解决代码BUG，解决后往往是恍然大悟，又是格外欣喜。如果当初陷入了一个死胡同里，可能就一直在那里徘徊，可见试错体验多么的刻骨铭心，这样足可以理解"量产经验"的重要性。

每一次的披荆斩棘，都是一笔人生财富。

5.1 轻量化网络模型设计

轻量化网络模型设计（Model Lightweight）是基于目标部署平台去设计更加高效的网络模型结构，其高效主要体现在减小参数量、减少计算量等方面。一般把部署在端侧的小模型称为轻量化模型。轻量化网络模型的开发方法如下：

首先，设计和训练一个大模型，这里的大，不是特指那些具有上亿数量级参数的网络模型，而是以满足需求为目标，在先不考虑硬件平台的情况下，把精力放在设计和训练一个精度满足要求的模型。然后主要的工作便是将模型压缩到满足在边缘端、移动端或嵌入式设备中运行的需求。下文主要以基本算子和基本模块去讨论如何去搭建和评估一个轻量化模型。设计模型和搭积木一样有趣，每一块积木，就是本节中讨论的算子和基本模块。

5.1.1 下采样

下采样（Downsampling）也叫降采样，是实现特征图从高分辨率到低分辨率的转变过程，如图5-1所示。

在深度学习中，下采样的方法主要是池化（Pooling）。池化的作用主要有以下几点：

（1）在保留主要特征的前提下减少模型的参数量，对防止模型过拟合起到一定的作用。

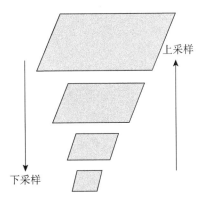

图 5-1 下采样、上采样示意图

（2）降低模型推理的计算量。

（3）感受野（Receptive Field）是指特征图中每一个像素点映射到原始输入图像上的区域大小。通过池化后输出的特征图变小，所以增大了输出特征图中每个像素点的感受野。大感受野能够增强网络模型的全局特征的提取能力，比如形状和轮廓。相反，小感受野用于提取特征图中的局部特征，比如边缘和纹理。

（4）对于平移旋转不变性，起到一定的作用。

池化可分为均值池化和最值池化等。

均值池化（Mean Pooling）是将输入特征图中每个滑动窗口区域的平均值作为输出特征图的值。图 5-2 所示为以下所示代码的具体的计算示意图，其中卷积核大小为（2，2），池化窗口移动步长为（2，2），padding='VALID'无填充，池化后的输出尺寸为（2，2）。当池化窗口的移动步长小于池化窗口时，相邻的池化窗口会有部分重叠，这样的池化操作作为重叠池化（Overlapping Pooling）。本书中为了便于读者朋友理解，尽量采用图示法讲解，尽量少用公式去计算。

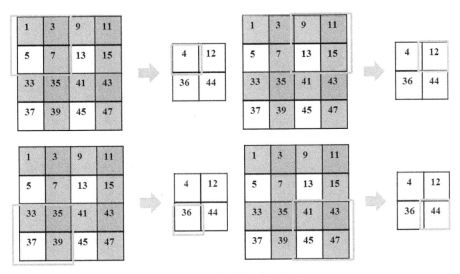

图 5-2 均值池化运算示意图

关于图 5-2 所示的均值池化的演示代码如下：

```python
import numpy as np
import tensorflow as tf
from tensorflow.keras.layers import AveragePooling2D, Reshape

x_ = [1, 3, 9, 11, 5, 7, 13, 15, 33, 35, 41, 43, 37, 39, 45, 47]
x_numpy = np.array(x_).reshape(1, -1)
# 这里的 reshape 是必须要加的。如果不加这一步操作，后面的操作会报错，读者朋友可以自行验证一下
print(x_numpy.shape) #(1, 16)

x_tensor = tf.convert_to_tensor(x_numpy, dtype=tf.float16)#numpy 转 tensor
print(x_tensor.shape)#(1, 16)

x = Reshape((4, 4, 1))(x_tensor)
print(x.shape) #(1, 4, 4, 1)

y = AveragePooling2D(pool_size=(2, 2), strides=(2, 2), padding='VALID',
                    data_format=None)(x)
print(y, y.shape)
```

代码运行结果：

```
tf.Tensor(
[[[[ 4.]
   [12.]]

  [[36.]
   [44.]]]], shape=(1, 2, 2, 1), dtype=float32)
```

其中 pool_size 为池化窗口大小，strides 为池化窗口移动步长，默认等于 pool_size。

关于以上验证代码，笔者在一开始代码测试中遇到以下一个问题：

Value for attr 'T' of int64 is not in the list of allowed values: half, bfloat16, float, double

出现这一问题的原因如下：

由于 NumPy 的默认数据类型为 int64，而 Tensor 目前不支持，所以解决方法如下：

将 x_tensor = tf.convert_to_tensor(x_numpy) 这行代码改为 x_tensor = tf.convert_to_tensor(x_numpy, dtype=tf.float16)。

如果增大池化窗的大小，结果会如何呢？如图 5-3 所示，我们将池化窗改为 (3, 3)，关键测试代码如下：

y = AveragePooling2D（pool_size =（3，3），strides =（2，2），padding='VALID'，data_format=None）(x)

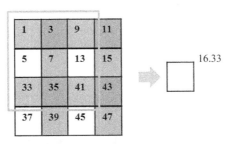

```
>>> y = AveragePooling2D(pool_size=(3, 3), strides=(2, 2), padding='VALID', data_format=None)(x)
>>> y
<tf.Tensor: shape=(1, 1, 1, 1), dtype=float32, numpy=array([[[[16.333334]]]], dtype=float32)>
>>> y.shape
TensorShape([1, 1, 1, 1])
>>>
```

图 5-3　池化窗大小为（3，3）的均值池化运算示意图和代码运行结果
[其中池化窗移动步长为（2，2），'VALID'无填充]

从图 5-3 中可以看出，x 的右边缘和下边缘数据未参加计算，也就是这些信息丢失掉，但是实际中参加计算的 x 尺寸很大，所以丢失的信息对一般模型任务影响不大。当然如果是图像超分任务，就需要考虑使用 padding = 'SAME'。当 padding = 'SAME' 时，参加运算的张量 x 的右边缘和下边缘会被填充数据 0，如图 5-4 所示，需要注意的一点是如果池化窗中有 padding 的 0，那么计算池化窗中的均值时，分母中不计入填充 0 的个数。在图中的右下角池化窗计算中是除以 4，而不是 9。这样做的目的是让输出特征图的尺寸满足实际中的要求和不丢失参加运算的输入特征图信息。

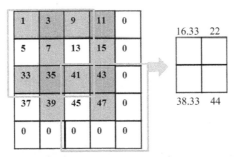

```
>>> y = AveragePooling2D(pool_size=(3, 3), strides=(2, 2), padding='SAME', data_format=None)(x)
>>> y
<tf.Tensor: shape=(1, 2, 2, 1), dtype=float32, numpy=
array([[[[16.333334],
         [22.      ]],

        [[38.333332],
         [44.      ]]]], dtype=float32)>
>>> y.shape
TensorShape([1, 2, 2, 1])
>>>
```

图 5-4　padding='SAME'下均值池化运算示意图和代码运行结果

最值池化（Max Pooling）是将输入特征图中每个滑动窗口区域的最大值作为输出特征图的值。图 5-5 所示为以下代码的运算示意图和运行结果，其中池化窗大小为（2，2），池化窗移动步长为（2，2），padding='VALID'无填充。

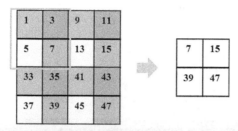

```
>>> y = MaxPool2D(pool_size=(2, 2), strides=(2, 2), padding='VALID', data_format=None)(x)
>>> y
<tf.Tensor: shape=(1, 2, 2, 1), dtype=float32, numpy=
array([[[[ 7.],
         [15.]],

        [[39.],
         [47.]]]], dtype=float32)>
>>> y.shape
TensorShape([1, 2, 2, 1])
```

图 5-5　最值池化运算示意图和代码运行结果

关于图 5-5 中的运算的演示代码如下：

import numpy as np

import tensorflow as tf

from tensorflow.keras.layers import Reshape，MaxPool2D

x_=[1，3，9，11，5，7，13，15，33，35，41，43，37，39，45，47]

x_numpy = np.array（x_）.reshape（1，-1）

print（x_numpy.shape）

x_tensor = tf.convert_to_tensor（x_numpy，dtype = tf.float16）

print（x_tensor.shape）

x = Reshape（(4，4，1)）(x_tensor）

print（x.shape）

y = MaxPool2D（pool_size =（2，2），strides =（2，2），padding =' VALID'，data_format = None）(x）#AveragePooling2D

print（y，y.shape）

代码运行结果：

tf.Tensor（

```
[[[[ 7.]
   [15.]]

  [[39.]
   [47.]]]], shape = (1, 2, 2, 1), dtype = float32)
```

其中 pool_size 为池化窗口大小，strides 为池化窗口移动步长，默认等于 pool_size。将以上演示代码中的池化窗口大小改为（3，3），得到的测试结果如图 5-6 所示。

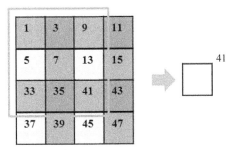

```
>>> y = MaxPool2D(pool_size=(3, 3), strides=(2, 2), padding='VALID', data_format=None)(x)
>>> y
<tf.Tensor: shape=(1, 1, 1, 1), dtype=float32, numpy=array([[[[41.]]]], dtype=float32>
>>> y.shape
TensorShape([1, 1, 1, 1])
>>>
```

图 5-6 池化窗大小为（3，3）时，最值池化运算示意图和代码运行结果

将以上测试代码中的 y = MaxPool2D（pool_size =（3，3），strides =（2，2），padding ='VALID'，data_format = None）(x) 的 padding 改为 'SAME'，测试结果如图 5-7 所示。

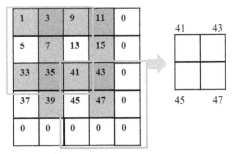

```
>>> y = MaxPool2D(pool_size=(3, 3), strides=(2, 2), padding='SAME', data_format=None)(x)
>>> y
<tf.Tensor: shape=(1, 2, 2, 1), dtype=float32, numpy=
array([[[[41.],
         [43.]],
        [[45.],
         [47.]]]], dtype=float32)>
>>> y.shape
TensorShape([1, 2, 2, 1])
>>>
```

图 5-7 当 padding='SAME' 时，最值池化运算示意图和代码运行结果

通过以上篇幅的讨论，相信读者朋友对下采样和池化操作有较深入的理解，接下来我们讨论上采样。

5.1.2 上采样

上采样（Upsampling）是实现特征图从低分辨率到高分辨率的转化过程，实现的方法有插值、PixelShuffle/depth_to_space、反池化、反卷积等。接下来讨论这几种方法的实现原理。

（1）Bilinear

Bilinear 是双线性插值算法，非常出名，建议读者朋友参考相关文献做深入理解。插值算法还有最近邻域插值（Nearest neighbor interpolation）、双三次插值或三次样条插值（Bicubic interpolation）等[①]，它们可以很方便地实现输入图像或特征图的放大和缩小，可用于上采样。

（2）UpSampling2D

UpSampling2D 是实现上采样的一个算子。它在 Keras 框架中的实现方法为：tf.keras.layers.UpSampling2D（size =（2，2），data_format = None，interpolation = 'nearest'，**kwargs），其中 size 为 rows 和 columns 方向的上采样因子，且为整数。UpSampling2D 的底层采用最近邻插值或双线性插值等插值算法。

对于测试代码：

y = tf.keras.layers.UpSampling2D（size =（2，2））(x)

当输入特征图维度为（1，4，4，1），得到的输出特征图维度为（1，8，8，1），如图 5-8（a）所示。

对于测试代码：

y = tf.keras.layers.UpSampling2D（size =（3，3））(x)

当输入特征图维度为（1，4，4，1），得到输出特征图维度为（1，12，12，1），如图 5-8（b）所示。

（3）PixelShuffle/depth_to_space

PixelShuffle 算子也叫亚像素卷积（Sub-Pixel Convolutional Neural Network），实现将维度为 $B \times H \times W \times (C \cdot r^2)$ 的低分辨率特征图通过像素重组得到维度为 $B \times (H \cdot r) \times (W \cdot r) \times C$ 的高分辨率特征图的上采样过程[②]。该算子在 Torch 框架中通过 torch.nn.PixelShuffle

---

[①] 可以参考学习 Geometric Image Transformations：
https://docs.opencv.org/5.x/da/d54/group__imgproc__transform.html#gga5bb5a1fea74ea38e1a5445ca803ff121ac97d8e4880d8b5d509e96825c7522deb

[②] Shi W, Caballero J, Huszár, Ferenc, et al.Real-Time Single Image and Video Super-Resolution Using an Efficient Sub-Pixel Convolutional Neural Network.arXiv e-prints, 2016.DOI:10.1109/CVPR.2016.207.

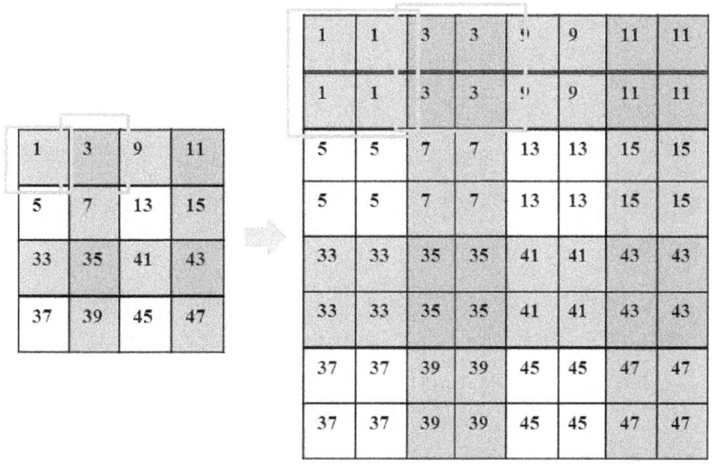

```
>>> x.shape
TensorShape([1, 4, 4, 1])
>>> y = tf.keras.layers.UpSampling2D(size=(2, 2))(x)
>>> y.shape
TensorShape([1, 8, 8, 1])
```

(a)

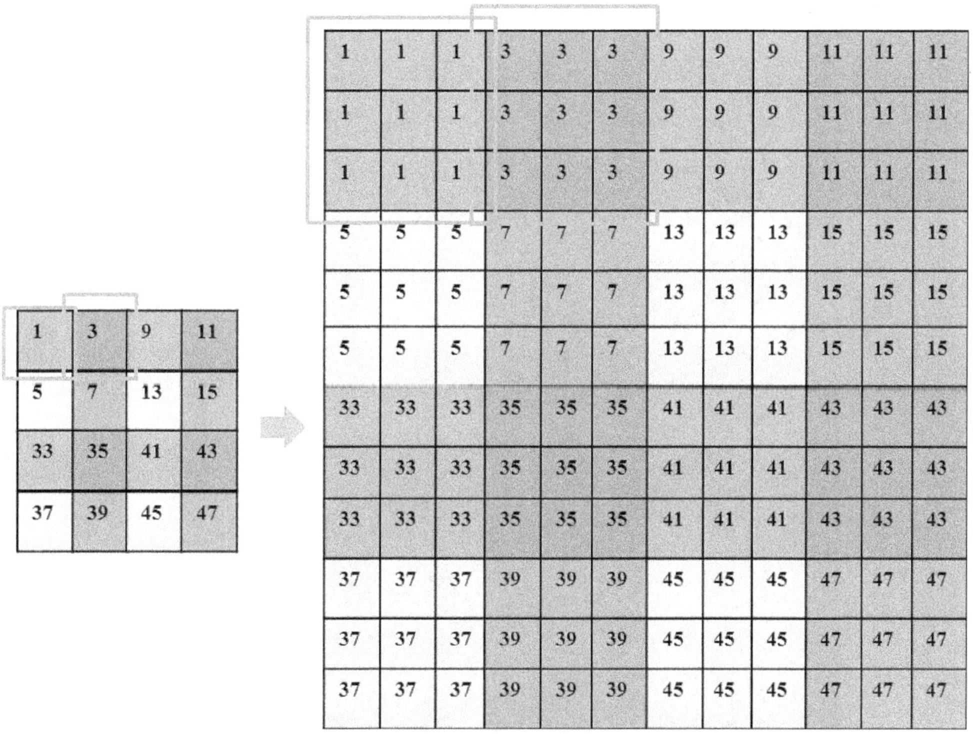

```
>>> y = tf.keras.layers.UpSampling2D(size=(3, 3))(x)
>>> y.shape
TensorShape([1, 12, 12, 1])
```

(b)

图 5-8　UpSampling2D 算子运算示意图和代码运行结果

实现，在 TensorFlow 框架中对应的算子名字是 depth_to_space。PixelShuffle（或 depth_to_space，下同）实现在低分辨率特征图做卷积、池化、激活、注意力机制等一系列算子运算，然后通过像素组合得到高分辨率特征图，这样的好处是减小模型计算量。对比前面介绍的插值算法，PixelShuffle 可以更好地保留图像的细节和纹理信息，具有更好的图像重建质量和更优的计算性能，非常适合图像超分、图像增强、去噪、去模糊、语义分割、图像融合等场景。

在 TensorFlow 框架中，数据存储默认采用 NHWC 格式。这种格式的数据将一个张量（特征图以张量的形式存储）中 C 个通道对应的空间位置的数据存储在一段连续的内存中。如图 5-9 所示的一段内存中，1—8 为 8 个通道相同位置的数据，9—16 同理。

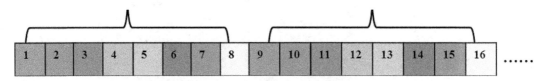

图 5-9　NHWC 格式存储

图 5-10 所示是 PixelShuffle 具体的实现流程。其中图（a）是尺寸为 4×4，通道数为 8 的张量（其中的内存排布如图 5-9 所示）。图（b）是先沿 W 方向，再沿 H 方向依次取出每个通道对应位置的数据，得到一个行张量，然后通过 reshape 操作实现张量形状的改变。图（c）是将 reshape 处理后得到的数据拷贝重新组合后得到一个新的张量。

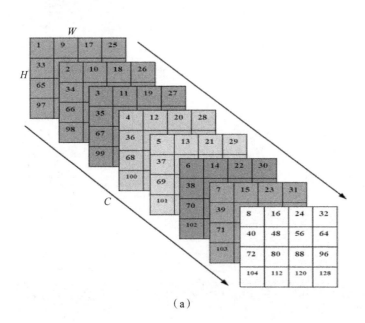

（a）

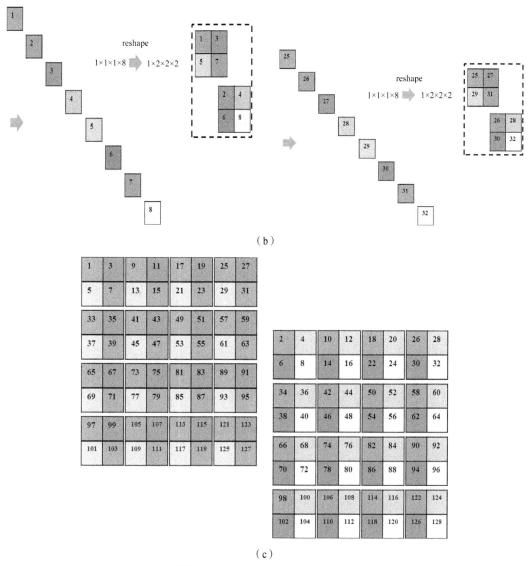

图 5-10 PixelShuffle 运算示意图

PixelShuffle 实现了分辨率增加，但是通道数减小了。

以上操作中用到的 reshape 只能改变张量的形状，不改变张量的物理存储顺序。关于 reshape 运算我们通过以下验证代码来进一步讨论。

```
import numpy as np
import tensorflow as tf
from tensorflow.keras.layers import Reshape

x_=[1, 3, 9, 11, 5, 7, 13, 15, 33, 35, 41, 43, 37, 39, 45, 47]
x_numpy = np.array（x_）.reshape（1, -1）
```

```
print(x_numpy.shape)

x_tensor = tf.convert_to_tensor(x_numpy, dtype=tf.float16)
print(x_tensor.shape)

x = Reshape((4, 4, 1))(x_tensor)
print(x.shape)

y = Reshape((2, 2, 4))(x)
print(y, y.shape)
```

代码运行得到的结果如下：

```
tf.Tensor(
[[[[ 1. 3. 9. 11.]
   [ 5. 7. 13. 15.]]

  [[33. 35. 41. 43.]
   [37. 39. 45. 47.]]]], shape=(1, 2, 2, 4), dtype=float32)
```

图 5-11 所示为以上代码中 reshape 的操作流程示意图，我们可以看到 reshape 后的存储顺序（BHWC 格式）始终没有改变，只是改变了形状（从 4×4×1 改为 2×2×4）。

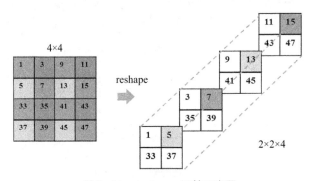

图 5-11　reshape 运算示意图

再次强调一下：BHWC 格式的张量是多个通道中对应空间位置的像素连续存储。而 NCHW 格式的张量是每个通道内的像素连续存储。

对于不同的硬件平台和不同的算子，往往不同的存储格式下推理性能会有很大的差异。如图 5-12 所示，在 NVIDIA A100-SXM4-80GB，CUDA 11.2，cuDNN 8.1 下，对于 transpose 算子，NHWC 格式存储的 TFLOPS 会优于 NCHW 格式存储的。

如果是池化算子，由于是按照同一通道的数值内存排布来计算的，NCHW 格式存储的 TFLOPS 一般会优于 NHWC 格式存储的。

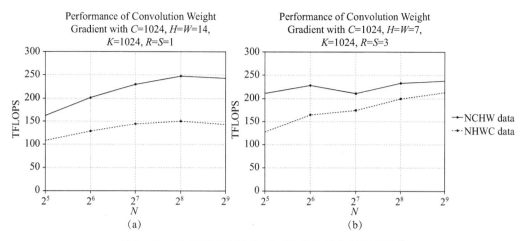

图 5-12　NHWC 和 NCHW 格式存储的 TFLOPS 评估[1]

对于其他硬件，同样存在性能的差异。

所以，在端侧模型部署中，我们需要结合硬件平台、算子特性去综合评估。

（4）反池化

反池化（Unpooling）目前在深度学习框架中没有实现算子，应用不多。鉴于本书篇幅限制，不做具体讨论，读者朋友可以参考相关文献自行研究。

（5）反卷积

反卷积（Transposed Convolution，转置卷积）是一种特殊的正向卷积运算[2]，用于实现特征图的上采样。先对输入特征图的内部元素之间填充 0 和边界补 0，实现放大特征图的尺寸，再做卷积运算（图 5-13）。

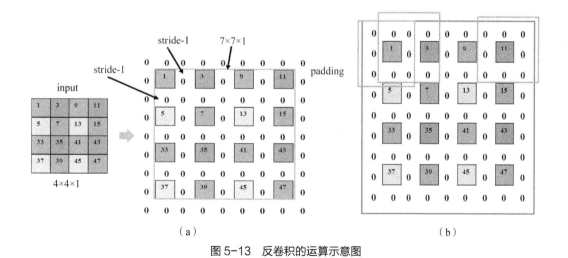

图 5-13　反卷积的运算示意图

---

[1] 图来源：https://docs.nvidia.com/deeplearning/performance/dl-performance-convolutional/index.html#imp-gemm-dim
[2] 更多可参考论文 Deconvolutional Networks，https://www.matthewzeiler.com/mattzeiler/deconvolutionalnetworks.pdf

图 5-13 所示的反卷积的运算流程中，输入特征图尺寸 4×4，卷积核尺寸为 3×3，步长为 1，填充为 1，也就是 $i=4$，$k=3$，$s=1$，$p=1$，那么反卷积后的尺寸为多少呢？

首先按照卷积的计算公式计算正常卷积后的输出尺寸大小：

$o=(i+2p-k)/s+1=(4+2\times1-3)/1+1=4$

反卷积的输出尺寸的计算分为两种情况：

①当 $(o+2p-k)\%s==0$：

反卷积的输出尺寸 output$=s(i-1)-2p+k$。

图 5-13 的 $(o+2p-k)\%s=(4+2-3)\%1=3$ 不满足 $==0$ 的条件。

②当 $(o+2p-k)\%s\neq0$：

反卷积的输出尺寸 output$=s(i-1)-2p+k+(o+2p-k)\%s$。

图 5-13 的 $(o+2p-k)\%s=3\neq0$，

所以反卷积的输出尺寸 output$=s(i-1)-2p+k+(o+2p-k)\%s=1\times(4-1)-2\times1+3+(o+2p-k)\%s=4+3=7$。

反卷积不同于普通的插值算法，它通过卷积核和输入特征图的卷积运算实现上采样，其中的卷积核参数是通过训练学习得到，所以上采样效果一般（之所以说一般，是因为不同的场景，往往模型的效果会有差异。我们在实际模型搭建和训练中，需要对训练得到的模型进行评估，然后做出最佳的上采样的方法选取）会比普通的插值算法更好。[①] 以下是 TensorFlow 框架中的反卷积的实现方法，其中 kernel_size 为卷积核的大小，strides 不是卷积核移动步长，而是在输入特征图中元素之间补充 strides-1 个 0，且反卷积的步长仍然为 1。

反卷积的输出维度等于卷积的输入维度。

反卷积的代码演示如下：

```
import tensorflow as tf
from tensorflow.keras.layers import Conv2DTranspose
x = tf.ones ( shape=[1, 640, 320, 3])
y = Conv2DTranspose ( filters=16, kernel_size=(4, 4), strides=(2, 2), padding=
              'same', use_bias=False )( x )
```

其中输入 x 的维度为 TensorShape ([1, 640, 320, 3])，得到的输出结果维度为 TensorShape ([1, 1280, 640, 16])。

5.1.3 全局池化

前文讨论的池化，通过池化窗口的滑动实现窗口区域的值的均值或最值操作，是一

---

① Dumoulin V, Visin F. A guide to convolution arithmetic for deep learning. 2016.DOI:10.48550/arXiv.1603.07285.

种局部池化。当池化窗口的尺寸等于特征图的尺寸时，池化窗对整个特征图区域做均值或最值池化，这种池化就是本小节讨论的全局池化。

全局池化是一种特殊的池化，可分为全局平均池化（Global Average Pooling，GAP）和全局最大池化。全局池化将尺寸为 $H \times W \times C$ 的张量通过平均或最值计算降维到尺寸大小为 $1 \times 1 \times C$ 的张量，如图 5-14 所示。

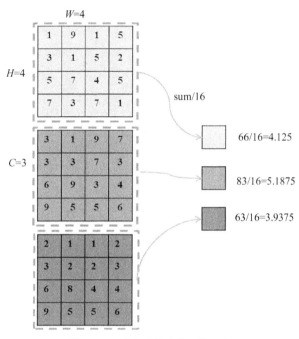

图 5-14　全局平均池化运算示意图

全局平均池化运算的代码演示如下：

import numpy as np

import tensorflow as tf

x_ =［1, 3, 2, 9, 1, 1, 1, 9, 1, 5, 7, 2,

　　3, 3, 3, 1, 3, 2, 5, 7, 2, 2, 3, 3,

　　5, 6, 6, 7, 9, 8, 4, 3, 4, 5, 4, 4,

　　7, 9, 9, 3, 5, 5, 7, 5, 5, 1, 6, 6］

x_numpy = np.array（x_）.reshape（1, -1）

print（x_numpy.shape）

x_tensor = tf.convert_to_tensor（x_numpy, dtype = tf.float16）

print（x_tensor.shape）

```
x = tf.keras.layers.Reshape((4, 4, 3))(x_tensor)
print(x, x.shape)

y = tf.keras.layers.GlobalAveragePooling2D()(x)
print(y, y.shape)
```

代码运行结果如下：

```
(1, 48)
tf.Tensor(
[[[[1. 3. 2.]
   [9. 1. 1.]
   [1. 9. 1.]
   [5. 7. 2.]]

  [[3. 3. 3.]
   [1. 3. 2.]
   [5. 7. 2.]
   [2. 3. 3.]]

  [[5. 6. 6.]
   [7. 9. 8.]
   [4. 3. 4.]
   [5. 4. 4.]]

  [[7. 9. 9.]
   [3. 5. 5.]
   [7. 5. 5.]
   [1. 6. 6.]]]], shape=(1, 4, 4, 3), dtype=float32)(1, 4, 4, 3)
tf.Tensor([[4.125 5.1875 3.9375]], shape=(1, 3), dtype=float32)(1, 3)
```

全局池化的好处如下：

① 显著降低模型的参数量，防止过拟合；

② 实现对全局特征的感知；

③ 代替全连接层[①]。

关于第③点可以通过图 5-15 和图 5-16 来进一步解释：

图 5-15 中的特征图通过 Flatten 展平后，再拼接为一个一维张量，然后输入到全

---

① 参考：Lin M, Chen Q, Yan S. Network In Network.2013.DOI:10.48550/arXiv.1312.4400.

连接层，再将全连接层［在 Pytorch 框架中全连接层的实现为 nn.Linear（ ）］的输出输入到 Softmax 层，得到的输出为概率分布。通过这个模型可以实现分类、回归等任务。图 5-15 中的输入张量通道 C 等于 6（这里的通道数不一定需要等于 6，可以通过后面的全连接层调整为 6），通过全连接后，输出的张量维度 k 等于 6，通过 Softmax 层后得到 6 个概率值（总和等于 1），代表每一类的概率。图中的输入张量通过展平和拼接后得到的一维张量，其长度等于 $H \times W \times C$，再输入到全连接层，在全连接层中的权重参数量等于 $H \times W \times C \times k$，可见非常大。该如何优化呢？

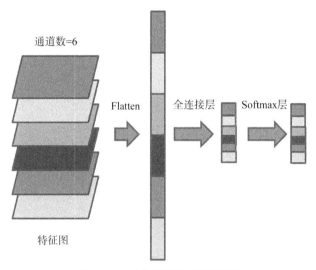

图 5-15　全连接层搭建模型网络

图 5-16 中将图 5-15 中的 Flatten 层、全连接层替换为全局池化层，这样参数量显著减小，可以有效防止过拟合和提高计算性能。

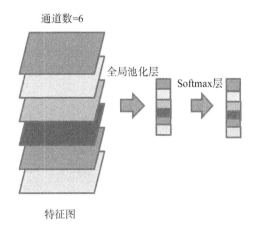

图 5-16　全局池化层搭建模型网络

以上讨论的全局池化是分别对每个通道的特征图通过最值或者均值计算得到一个值。如果我们是将整个通道在每个特征图上对应位置的像素做最值或者均值计算，最终输出一个 $H \times W$ 的张量，这便是全局深度最大/均值池化，如图 5-17 所示。

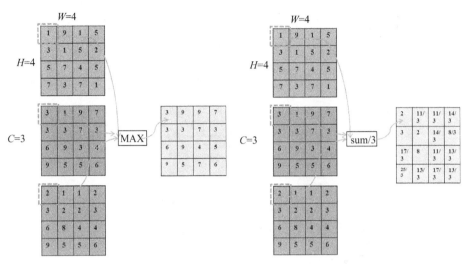

图 5-17　全局深度最大/均值池化运算示意图

5.1.4　分组卷积

什么是分组卷积（Group Convolution）？在这里通过图 5-18 给读者朋友介绍其运算的原理。

如图 5-18 所示，输入特征图的通道数等于 9（记为 C），卷积核数量等于 6（记为 N）。如果分组数为 3（记为 G），那么将卷积核分组后的计算情况如下：

每组的卷积核数 $=N/G=2$；

每组的输入特征图数 $=C/G=3$；

每个卷积核的通道数 $=C/G=3$。

图 5-18 所示的分组卷积代码演示如下（基于 TensorFlow 框架）：

```
import numpy as np
import tensorflow as tf

x_ = np.arange（4*4*9）
x_numpy = np.array（x_）.reshape（1，-1）
print（x_numpy.shape）

x_tensor = tf.convert_to_tensor（x_numpy，dtype = tf.float16）
print（x_tensor.shape）
```

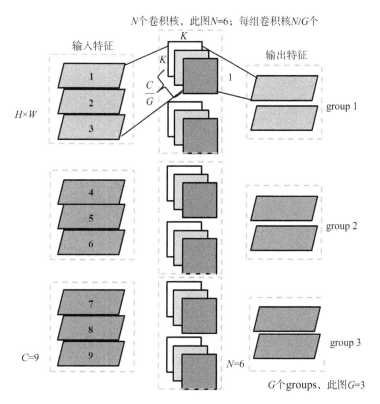

图 5-18 分组卷积运算示意图

```
x = tf.keras.layers.Reshape((4, 4, 9))(x_tensor)
print(x, x.shape) # (1, 4, 4, 9)

y = tf.keras.layers.Conv2D(6, kernel_size=(3, 3),
        strides=(2, 2), padding='SAME', groups=3)(x)
# 其中输入通道数 C 等于 9，卷积核数量 N 等于 6，分组数量 G 等于 3
print(y, y.shape) # (1, 2, 2, 6)
```

分组卷积的好处：

参加运算的卷积核维度从 $K \times K \times C$ 变为 $K \times K \times (C/G)$，即卷积核通道数变为原来的 $1/G$，所以整体参数量是原来的 $1/G$，从而实现计算性能的优化。

5.1.5 全局加权池化

全局加权池化（Global Depthwise Convolution，GDC）和全局平均池化不同的是，每个通道的特征图中的每一个位置的像素都乘以其对应卷积核中的权重参数，再求和得到一个数值，如图 5-19 所示。

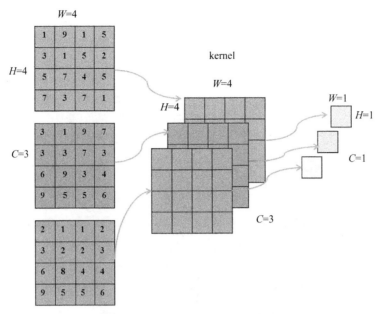

图 5-19　全局加权池化运算示意图

全局加权池化的代码演示如下：

```
import numpy as np
import tensorflow as tf

x_ =[1, 3, 2, 9, 1, 1, 1, 9, 1, 5, 7, 2,
     3, 3, 3, 1, 3, 2, 5, 7, 2, 2, 3, 3,
     5, 6, 6, 7, 9, 8, 4, 3, 4, 5, 4, 4,
     7, 9, 9, 3, 5, 5, 7, 5, 5, 1, 6, 6]
x_numpy = np.array（x_）.reshape（1, -1）
print（x_numpy.shape）

x_tensor = tf.convert_to_tensor（x_numpy, dtype = tf.float16）
print（x_tensor.shape）

x = tf.keras.layers.Reshape（(4, 4, 3)）（x_tensor）
print（x, x.shape）

y = tf.keras.layers.DepthwiseConv2D（kernel_size =（x.shape [1], x.shape [2]）, strides =（1, 1）, padding ='VALID'）（x）
print（y, y.shape）
```

演示代码输出结果如下：

（1，48）

tf.Tensor（

[[[[1. 3. 2.]

[9. 1. 1.]

[1. 9. 1.]

[5. 7. 2.]]

[[3. 3. 3.]

[1. 3. 2.]

[5. 7. 2.]

[2. 3. 3.]]

[[5. 6. 6.]

[7. 9. 8.]

[4. 3. 4.]

[5. 4. 4.]]

[[7. 9. 9.]

[3. 5. 5.]

[7. 5. 5.]

[1. 6. 6.]]]]，shape =（1，4，4，3），dtype = float32）（1，4，4，3）

tf.Tensor（[[[[3.2055697 5.250785 −5.8751163]]]]，shape =（1，1，1，3），dtype = float32）（1，1，1，3）

5.1.6 1×1 卷积

我们知道卷积核的尺寸为 $H \times W$，当 $H=W=1$ 的时候，这个卷积便叫作 1×1 卷积。

1×1 卷积的作用：

（1）实现跨通道信息融合。

如图 5-20 所示，1×1 卷积实现通道信息融合。图中为 1 个 1×1 卷积核，且通道数等于 3，和输入特征图的通道数相等。

（2）通过改变参加 1×1 卷积的卷积核的数量，实现输出特征图通道数的调整。

以下演示代码中，输入特征图维度为 TensorShape（[2，640，320，32]）。参数 kernel_size =（1，1），表明这个卷积为 1×1 卷积。将参数 filters 设置为 64，表明卷积核的数量为 64。最终卷积运算后的输出特征图的通道数等于 64，也就是实现了通道数的调整。

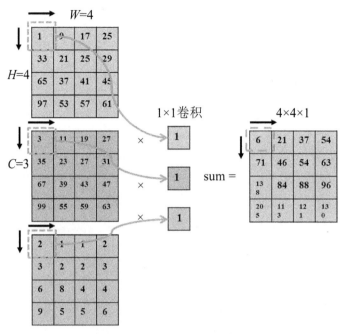

图 5-20　1×1 卷积实现通道信息融合

```
import tensorflow as tf
x = tf.random.normal（[2, 640, 320, 32]）
conv_1 = tf.keras.layers.Conv2D（64, kernel_size=（1, 1）, strides=（1, 1）, padding ='SAME'）
y = conv_1（x）
```

测试得到 y.shape 等于 TensorShape（[2, 640, 320, 64]）。

（3）通过卷积运算中添加的激活函数来增加模型网络的非线性特征。

1×1 卷积代码演示如下（基于 TensorFlow 框架）：

```
import numpy as np
import tensorflow as tf

x_ =[1, 3, 2, 9, 11, 1, 17, 19, 1, 25, 27, 2,
     33, 35, 3, 21, 23, 2, 25, 27, 2, 29, 31, 3,
     65, 67, 6, 37, 39, 8, 41, 43, 4, 45, 47, 4,
     97, 99, 9, 53, 55, 5, 57, 59, 5, 61, 63, 6]
x_numpy = np.array（x_）.reshape（1, -1）
print（x_numpy.shape）

x_tensor = tf.convert_to_tensor（x_numpy, dtype = tf.float16）
```

```
print ( x_tensor.shape )

x = tf.keras.layers.Reshape ( ( 4, 4, 3 ) )( x_tensor )
print ( x, x.shape )

y = tf.keras.layers.Conv2D ( 1, kernel_size = ( 1, 1 ), strides = ( 1, 1 ),
            padding = ′ SAME ′, kernel_initializer = ′ ones ′ )( x )
print ( y, y.shape )
```

代码验证结果如下：

tf.Tensor（

[[[[1. 3. 2.]

　[9. 11. 1.]

　[17. 19. 1.]

　[25. 27. 2.]]

　[[33. 35. 3.]

　[21. 23. 2.]

　[25. 27. 2.]

　[29. 31. 3.]]

　[[65. 67. 6.]

　[37. 39. 8.]

　[41. 43. 4.]

　[45. 47. 4.]]

　[[97. 99. 9.]

　[53. 55. 5.]

　[57. 59. 5.]

　[61. 63. 6.]]]], shape = (1, 4, 4, 3), dtype = float32)(1, 4, 4, 3)

tf.Tensor（

[[[[6.]

　[21.]

　[37.]

　[54.]]

　[[71.]

　[46.]

```
    [ 54.]
    [ 63.]]

  [[138.]
   [ 84.]
   [ 88.]
   [ 96.]]

  [[205.]
   [113.]
   [121.]
   [130.]]]], shape=(1, 4, 4, 1), dtype=float32)(1, 4, 4, 1)
```

1×1卷积应用在很多的模型网络中，比如Inception、YOLOv8、LKD-Net等。

LKD-Net（Large Kernel Convolution Network）模型的基本模块LKD模块中的DLKCB（Decomposition Large Kernel Convolution Block，分解大核卷积块）模块使用2次1×1卷积，用于增加多个通道信息的融合，如图5-21所示。

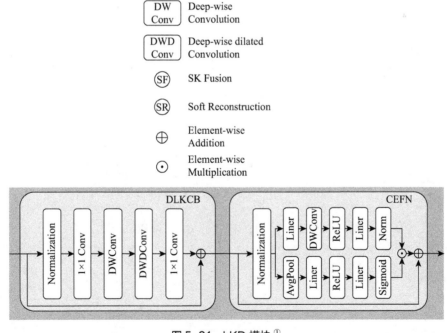

图 5-21 LKD 模块[1]

---

[1] 图来源：Luo Pinjun, Xiaoguo Qiang, Gao Xinbo, et al. LKD-Net: Large Kernel Convolution Network for Single Image Dehazing. In Proceedings of 2023 IEEE International Conference on Multimedia and Expo（ICME）2023..

以上的LKD模块的代码实现如下（基于PyTorch框架）[①]：

```python
import torch
import torch.nn as nn
class LKDBlock（nn.Module）:
    def __init__（self, network_depth, dim, mlp_ratio = 4.）:
        super（）.__init__（）

        # DLKCB
        self.norm1 = nn.BatchNorm2d（dim）
        self.Linear1 = nn.Conv2d（dim, dim, 1）
        self.DWConv = nn.Conv2d（dim, dim, 5, padding = 2,
                    groups = dim, padding_mode =' reflect'）
        self.DWDConv = nn.Conv2d（dim, dim, 7, stride = 1,
                    padding = 9, groups = dim, dilation = 3,
                    padding_mode=' reflect'）
        self.Linear2 = nn.Conv2d（dim, dim, 1）

        # CEFN
        self.norm2 = nn.BatchNorm2d（dim）
        self.cemlp = CEFN（network_depth = network_depth,
                    dim = dim, hidden_features = int（mlp_ratio）
                    * dim, out_features = dim）

    def forward（self, x）:
        identity = x
        x = self.norm1（x）
        x = self.Linear1（x）
        x = self.DWConv（x）
        x = self.DWDConv（x）
        x = self.Linear2（x）+ identity

        identity = x
        x = self.norm2（x）
        x = self.cemlp（x）+ identity
        return x
```

---

① 代码来源：https://github.com/SWU-CS-MediaLab/LKD-Net

5.1.7 深度卷积

深度卷积（Depthwise Convolution）实现输入特征图的每一个通道和一个单通道的卷积核的卷积运算，如图 5-22 所示。

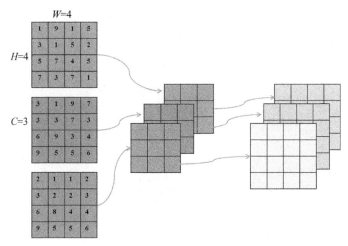

图 5-22　深度卷积运算示意图

深度卷积的代码演示如下：

```
y = tf.keras.layers.DepthwiseConv2D（kernel_size =（3，3），strides =（1，1），
                    padding ='SAME'）(x)
```

5.1.8　逐点卷积

逐点卷积（Pointwise Convolution）是卷积核大小为 1×1 的标准卷积。逐点卷积和 1×1 卷积的操作是相同的。

如图 5-23 所示，卷积核通道数（这里等于 3）等于输入特征图的通道数（这里等于 3）。

逐点卷积可通过以下代码实现［注意其中的 kernel_size=（1，1）］：

```
y = tf.keras.layers.Conv2D（32, kernel_size =（1，1），strides =（1，1），
                    padding ='SAME'）(x)
```

5.1.9　异构卷积

异构卷积（Heterogeneous Convolutions，HetConv）是在对输入特征图进行卷积运算时，其中部分特征图的通道采用 $k \times k$ 卷积核（如 3×3 卷积核），其余部分通道采用 1×1 卷积核，而且这两种卷积核在参加运算中不是先后关系，而是并行的。

如图 5-24 所示是异构卷积和标准卷积的差异，其中 M 是卷积核的数量，等于输入特征图通道的数量。P 等于卷积核数量 M 和其中为 3×3 卷积核数量的比值，则 3×3 卷

积核等于 M/P 个，1×1 卷积核等于 $M-M/P$ 个[①]。

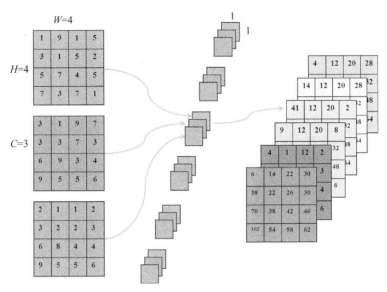

图 5-23 逐点卷积运算示意图

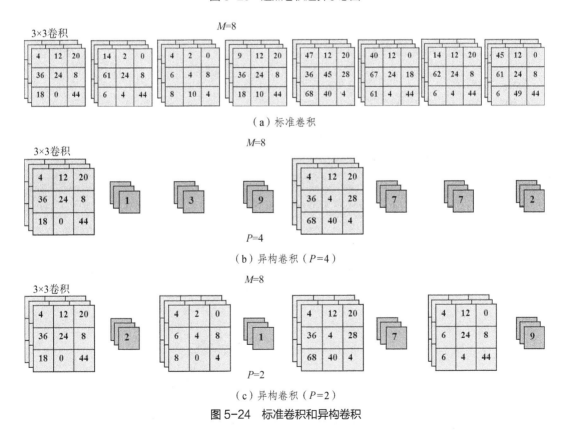

（a）标准卷积

（b）异构卷积（$P=4$）

（c）异构卷积（$P=2$）

图 5-24 标准卷积和异构卷积

---

① 参考 Singh P, Verma V K, Rai P, et al. HetConv: Heterogeneous Kernel-Based Convolutions for Deep CNNs. 2019.

异构卷积的演示代码如下[①]：

```
class HetConv（nn.Module）:
    def __init__（self, in_channels, out_channels, p）:
        super（HetConv, self）.__init__（）
        # Groupwise Convolution
        self.gwc = nn.Conv2d（in_channels, out_channels,
              kernel_size = 3, padding = 1, groups = p, bias = False）
        # Pointwise Convolution
        self.pwc = nn.Conv2d（in_channels, out_channels, kernel_size = 1, bias = False）

    def forward（self, x）:
        return self.gwc（x）+ self.pwc（x）
```

异构卷积的好处：

在保证模型精度的前提下，降低模型推理的计算成本。但是在实际端侧 DSP 核中，可能会增加计算成本。

5.1.10 深度可分离卷积

深度可分离卷积（Depthwise Separable Convolution）是深度卷积和逐点卷积两个网络模块的级联，如图 5-25 所示。

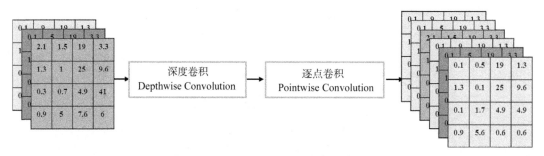

图 5-25 深度可分离卷积的模型网络结构

深度可分离卷积代码演示如下：

```
y1 = tf.keras.layers.DepthwiseConv2D（kernel_size =（3, 3），
        strides =（1, 1）, padding = 'SAME'）(x)# 深度卷积
y = tf.keras.layers.Conv2D（32, kernel_size =（1, 1），
        strides =（1, 1）, padding = 'SAME'）(y1)# 逐点卷积
```

[①] 代码来源：https://github.com/irvinxav/Efficient-HetConv-Heterogeneous-Kernel-Based-Convolutions

深度可分离卷积的计算效率远优于标准卷积，但是在很多模型中会出现精度的丢失，鱼和熊掌往往很难兼得。

5.1.11 空洞卷积

空洞卷积（也叫扩张卷积）的卷积核和普通卷积核不一样，它的卷积核中元素的横向和纵向之间位置补充 dilated-1 个零，所以空洞卷积的卷积核被扩大。其中 dilated 是卷积扩张率（dilation rate），它是用于控制扩大卷积核的程度。空洞卷积的好处：

在不采用下采样算子和计算参数量不增加的情况下，实现扩大卷积核的感受野，如图 5-26 所示。

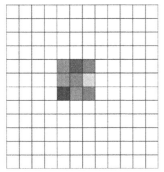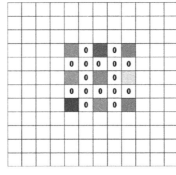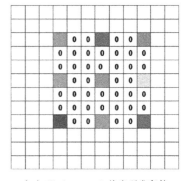

（a）普通卷积核，dilation rate=1　（b）dilation rate=2 的空洞卷积核　（c）dilation rate=3 的空洞卷积核

图 5-26　空洞卷积核和普通卷积核的区别

空洞卷积通过 dilation_rate 参数实现不同的扩张卷积，代码演示如下：

（1）空洞卷积，dilation_rate =（2，2）或（3，3）

```
y = layers.Conv2D ( filters, kernel_size, strides = ( 1, 1 ), padding =' SAME',
        dilation_rate = ( 2, 2 ), activation = None, use_bias = True )( x )
y = layers.Conv2D ( filters, kernel_size, strides = ( 1, 1 ), padding =' SAME',
        dilation_rate = ( 3, 3 ), activation = None, use_bias = True )( x )
```

（2）深度可分离空洞卷积，dilation_rate =（2，2）或（3，3）

```
y = layers.DepthwiseConv2D ( kernel_size = ( 3, 3 ), strides = ( 1, 1 ),
        dilation_rate = ( 2, 2 ), padding =' SAME' )( x )
y = layers.DepthwiseConv2D ( kernel_size = ( 7, 7 ), strides = ( 1, 1 ),
        dilation_rate = ( 3, 3 ), padding =' SAME' )( x )
```

如图 5-27 所示的 VAN（Visual Attention Network，视觉注意力网络）模块应用到空

洞卷积，图中的 d 代表卷积扩张率（dilation rate）[1]。

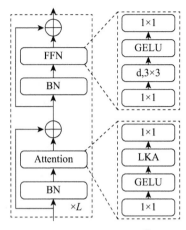

图 5-27 VAN 模块[2]

5.1.12 跳跃连接

跳跃连接（Skip Connections）将模型网络中某些层的输出传输到模型网络的更深层，实现跳过一些层连接。

如图 5-28 所示的 UNet 网络中使用 concatenate 拼接实现跳跃连接，实现浅层特征和深层特征的融合。

---

[1] 网络结构实现代码可以参考：https://github.com/Visual-Attention-Network/VAN-Classification/blob/main/models/van.py

[2] 图来源：Visual Attention Network

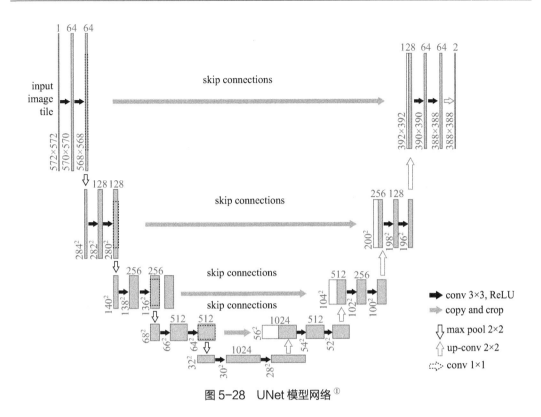

图 5-28　UNet 模型网络[①]

ResNet 模型的残差网络中的跳跃连接一般叫作残差连接（Residual Connection）。如图 5-29 所示，ResNet 网络中残差连接使用 add 相加实现，用于防止梯度爆炸和梯度消失。

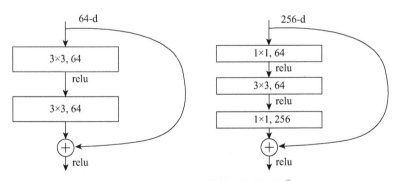

图 5-29　ResNet 模型中残差网络[②]

---

① 图来源：Ronneberger O, Fischer P, Brox T. U-Net: Convolutional Networks for Biomedical Image Segmentation. Springer International Publishing, 2015.DOI:10.1007/978-3-319-24574-4_28.

② 图来源：He K, Zhang X, Ren S, et al.Deep Residual Learning for Image Recognition. IEEE, 2016.DOI:10.1109/CVPR.2016.90.

如图 5-30 所示为 LKD-Net 模型的架构，可以看到用到很多的跳跃连接。

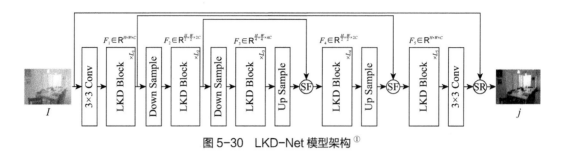

图 5-30　LKD-Net 模型架构[1]

如图 5-31 所示为 ShuffleNet 模型网络结构，用到跳跃连接。

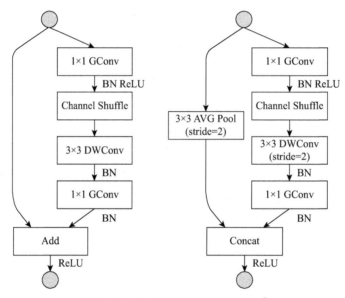

图 5-31　ShuffleNet 模型网络结构[2]

5.1.13　Flatten

Flatten 的功能是将一个维度为（B, H, W, C）的张量展平为维度为（$B, H \times W \times C$）。Flatten 代码演示如下：

```
import numpy as np
import tensorflow as tf
```

---

[1] 图来源：Luo PJ, Qiang XG, Gao XB, et al. LKD-Net: Large Kernel Convolution Network for Single Image Dehazing. In Proceedings of 2023 IEEE International Conference on Multimedia and Expo（ICME）2023.

[2] 图来源：Zhang X, Zhou X, Lin M, et al. Shufflenet: An extremely efficient convolutional neural network for mobile devices.Proceedings of the IEEE conference on Computer Vision and Pattern Recognition，2018: 6848-6856.

```
x_ = [1, 3, 9, 11, 5, 7, 13, 15, 33, 35, 41, 43, 37, 39, 45, 47]
x_numpy = np.array(x_).reshape(1, -1)
print(x_numpy.shape)

x_tensor = tf.convert_to_tensor(x_numpy, dtype=tf.float16)
print(x_tensor.shape)

x = tf.keras.layers.Reshape((4, 4, 1))(x_tensor)
print(x.shape)

y = tf.keras.layers.Flatten()(x)
print(y, y.shape)
```

代码运行结果：

(1, 16)
(1, 4, 4, 1)
tf.Tensor([[1. 3. 9. 11. 5. 7. 13. 15. 33. 35. 41. 43. 37. 39. 45. 47.]], shape=(1, 16), dtype=float32)(1, 16)

5.1.14 BatchNormalization

BN 层（BatchNormalization Layer，批标准化层）的计算步骤如下：[1]

（1）计算 $B \times H \times W \times C$ 个数据的均值；

（2）求数据的方差；

（3）数据标准化处理；

（4）模型训练，得到参数 γ 和 β；

（5）γ 和 β 线性变换得到输出值 y。

γ 和 β 是通过模型训练得到的参数，分别对应 BN 层的 weights 和 bias。

BN 层的演示代码如下：

```
import numpy as np
import tensorflow as tf

x_ = [1, 2, 3, 4, 5, 6, 7, 8, 9, 10, 11, 12, 13, 14, 15, 16]
x_numpy = np.array(x_).reshape(1, -1)
print(x_numpy.shape)
```

---

[1] 参考资料：Ioffe S, Szegedy C. Batch Normalization: Accelerating Deep Network Training by Reducing Internal Covariate Shift. JMLR.org, 2015.

```
x_tensor = tf.convert_to_tensor（x_numpy, dtype = tf.float16）
print（x_tensor.shape）

x = tf.keras.layers.Reshape（(4, 4, 1)）(x_tensor)
print（x, x.shape）

y = tf.keras.layers.BatchNormalization（）(x)
print（y, y.shape）
```

代码运行结果：

(1, 16)

tf.Tensor（

[[[[1.]

　　[2.]

　　[3.]

　　[4.]]

　[[5.]

　　[6.]

　　[7.]

　　[8.]]

　[[9.]

　　[10.]

　　[11.]

　　[12.]]

　[[13.]

　　[14.]

　　[15.]

　　[16.]]]], shape=(1, 4, 4, 1), dtype = float32）(1, 4, 4, 1)

tf.Tensor（

[[[[0.9995004]

　　[1.9990008]

　　[2.9985013]

　　[3.9980016]]

[[4.997502]

[5.9970026]

[6.996503]

[7.996003]]

[[8.995503]

[9.995004]

[10.994504]

[11.994005]]

[[12.9935055]

[13.993006]

[14.992506]

[15.992006]]]], shape =（1，4，4，1），dtype = float32)（1，4，4，1）

5.1.15 Dropout

Dropout 层实现模型在训练（Train）阶段，以概率 p 让某些神经元输出为 0，从而提高模型的泛化能力，防止模型过拟合。

在模型推理（Inference）阶段，Dropout 层一般处于关闭状态，如图 5-32 所示。

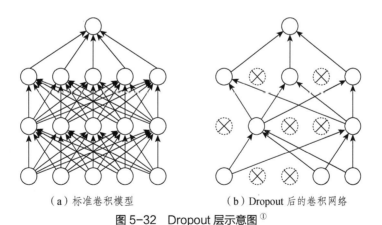

（a）标准卷积模型　　　　　（b）Dropout 后的卷积网络

图 5-32　Dropout 层示意图[①]

Dropout 层代码实现如下：

y = tf.keras.layers.Dropout（.2, input_shape =（2, ））(x)

---

① 图来源：Srivastava N, Hinton GE, Krizhevsky A, et al. Dropout: A Simple Way to Prevent Neural Networks from Overfitting. Journal of Machine Learning Research, 2014, 15（1）:1928-1958.

5.1.16 全连接层

全连接层（Full Connection Layer，FC）的每一个结点都连接上一层的结点，如图 5-33 所示。

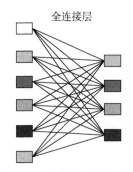

图 5-33　全连接层运算示意图

全连接层，在 Keras 框架中的算子名称叫 Dense 层，在 PyTorch 框架中的算子名称叫 Linear 层。其中 Dense 层代码演示如下：

```
import numpy as np
import tensorflow as tf

x_ =[1, 3, 9, 11, 5, 7, 13, 15, 33, 35, 41, 43, 37, 39, 45, 47]
x_numpy = np.array（x_）.reshape（1, -1）
print（x_numpy.shape）

x_tensor = tf.convert_to_tensor（x_numpy, dtype = tf.float16）
print（x_tensor.shape）

x = tf.keras.layers.Reshape（（16, 1））（x_tensor）
print（x.shape）#（1, 16, 1）

y = tf.keras.layers.Dense（5, activation =' relu'）（x）# 定义 5 个神经元的 Dense 层
print（y.shape）#（1, 16, 5）
```

5.1.17　SENet

SENet（Squeeze-and-Excitation Network，压缩和激励网络）是一种通道注意力机制模块（Channel Attention），不同于空间注意力（Spatial Attention）。

所谓通道注意力机制是指通过模型训练将输入特征图的多个通道中的作用大的通道的权重加大，而作用小的通道的权重减小。

SENet 的基本模块为 SE 模块。该模块通过自身和一个具备全局感受野的 $1\times1\times C$ 特征图相乘得到一个新的特征图。如图 5-34 所示，SE 模块包括压缩过程和激励过程。压缩过程是将输入特征图通过全局平均池化压缩为 $1\times1\times C$ 的特征图——这个特征图具备全局感受野的特征。激励过程为将压缩得到的 $1\times1\times C$ 特征图通过全连接层（FC）预测出代表每个通道的权重关系。

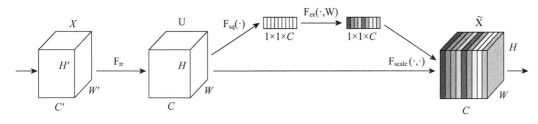

图 5-34　SE 模块网络结构[①]

5.1.18　MobileNet

MobileNet 是 2017 年 Google 提出的一种轻量化模型，它的第一个版本——MobileNet V1 创新性引入深度可分离卷积来构建轻量化模型，如图 5-35 所示。

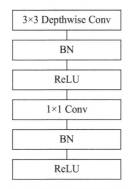

图 5-35　深度可分离卷积网络结构[②]

表 5-1 所示为 MobileNet V1 完整的模型网络结构，不计 Pool 层和 Softmax 层，共 28 层，其中 Conv 为标准卷积，Conv dw 为深度可分离卷积，s1 代表步长为 1，s2 代表步长为 2。

---

① 图来源：Hu J, Shen L, Sun G. Squeeze-and-Excitation Networks. 2018 IEEE/CVF Conference on Computer Vision and Pattern Recognition（CVPR）.IEEE，2018.DOI:10.1109/CVPR.2018.00745.

② 图来源：Howard A G, Zhu M, Chen B, et al. MobileNets: Efficient Convolutional Neural Networks for Mobile Vision Applications. 2017.

表 5-1 MobileNet V1 模型网络结构[1]

Type/Stride	Filter Shape	Input Size
Conv/s2	3×3×3×32	224×224×3
Conv dw/s1	3×3×32 dw	112×112×32
Conv/s1	1×1×32×64	112×112×32
Conv dw/s2	3×3×64 dw	112×112×64
Conv/s1	1×1×64×128	56×56×64
Conv dw/s1	3×3×128 dw	56×56×128
Conv/s1	1×1×128×128	56×56×128
Conv dw/s2	3×3×128 dw	56×56×128
Conv/s1	1×1×128×256	28×28×128
Conv dw/s1	3×3×256 dw	28×28×256
Conv/s1	1×1×256×256	28×28×256
Conv dw/s2	3×3×256 dw	28×28×256
Conv/s1	1×1×256×512	14×14×256
5× Conv dw/s1	3×3×512 dw	14×14×512
5× Conv/s1	1×1×512×512	14×14×512
Conv dw/s2	3×3×512 dw	14×14×512
Conv/s1	1×1×512×1024	7×7×512
Conv dw/s2	3×3×1024 dw	7×7×1024
Conv/s1	1×1×1024×1024	7×7×1024
Avg Pool/s1	Pool 7×7	7×7×1024
FC/s1	1024×1000	1×1×1024
Softmax/s1	Classifier	1×1×1000

MobileNet V2 在 MobileNet V1 基础上引入残差连接，如图 5-36 所示。

MobileNet V3（分为 MobileNetV3-Large 和 MobileNetV3-Small）在 MobileNets V2 的基础上引入 SE 模块[2]，以及将 ReLU 和 Sigmoid 激活函数分别改为 HardSwish 与 HardSigmoid。

---

[1] 表来源：Howard A G, Zhu M, Chen B, et al. MobileNets: Efficient Convolutional Neural Networks for Mobile Vision Applications. 2017.

[2] 参考文献：Howard A, Sandler M, Chu G，et al.Searching for MobileNetV3. 2019.DOI:10.48550/arXiv.1905.02244.

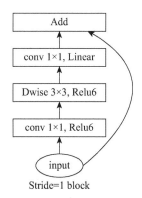

图 5-36 MobileNet V2 残差连接结构[1]

5.1.19 注意力机制

注意力机制是模仿人的眼睛在看物体时将视力聚焦在物体的某一处。在深度学习中通过模型的学习得到一组权重参数，用这些权重参数调整不同的特征的重要性，其中重要的特征权重大，反之不重要的特征权重小。注意力机制分为空间注意力（Spatial Attention Module）、通道注意力（Channel Attention Module，前文介绍的 SENet 模型属于这种类型）、自注意力（Self-Attention Module）等等。图 5-37 所示为 CBAM 注意力模块，由通道注意力模块和空间注意力模块组成。其中通道注意力模块用于增强不同通道的重要性，空间注意力模块增强特征图中不同空间位置的信息的重要性。通过通道注意力模块和空间注意力模块的组合使用，可以有效提高模型网络的特征提取能力。[2]

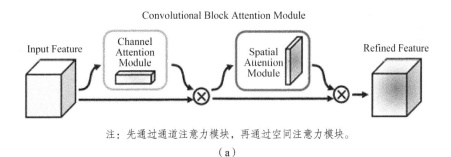

注：先通过通道注意力模块，再通过空间注意力模块。

（a）

---
[1] 图来源：MobileNet V2: Inverted Residuals and Linear Bottlenecks
[2] 参考实现代码：https://github.com/Jongchan/attention-module/blob/master/MODELS/cbam.py

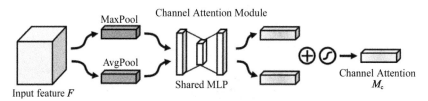

注：输入特征图经过全局最大池化和全局平均池化，得到表示每个通道的全局最大特征和平均特征；通过全连接层实现对以上两个全局特征的学习，得到通道的重要性权重；将两个特征重要性权重相加后通过 Sigmoid 函数输出值范围在 0-1 之间的权重参数；将通道权重参数与原先的输入特征图中每个通道相乘，得到能够体现出通道注意力信息的输出特征图。

（b）

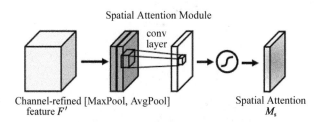

注：将输入特征图通过全局深度最大 / 均值池化（注意是按照通道维度）处理后，将输出特征合并为通道数为 2 的特征图，再分别通过卷积层和 Sigmoid 函数层，输出通道数为 1，值范围在 0-1 之间的权重参数；将空间权重参数与原先的输入特征图相乘，得到能够体现出空间注意力信息的输出特征图。

（c）

图 5-37　CBAM 注意力模块 [1]

自注意力机制是捕获输入序列的内部不同位置的元素的依赖关系。缩放点积注意力（Scaled Dot-Product Attention）是一种自注意力机制，它的运算流程如图 5-38 所示：

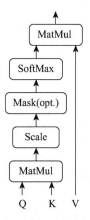

图 5-38　缩放点积注意力计算流程 [2]

输入 X 向量通过相互独立的线性变换 query、key、value 得到查询矩阵 Q、键矩阵 K、值矩阵 V 分量三个向量的初始。缩放点积注意力机制的计算公式如下：

---

[1] 图来源：Woo S, Park J, Lee J Y, et al. CBAM: Convolutional Block Attention Module. Springer, Cham, 2018. DOI:10.1007/978-3-030-01234-2_1.

[2] 图来源：Vaswani A, Shazeer N, Parmar N, et al. Attention Is All You Need. arXiv, 2017.

$$\text{Attention}(Q, K, V) = \left[\text{Softmax}(\frac{QK^T}{\sqrt{d_k}})\right]V, \text{其中 } d_k \text{ 为 } Q, K \text{ 列数} \quad (5.1\text{-}1)$$

图 5-39 更加具体地演示了以上公式的运算流程：

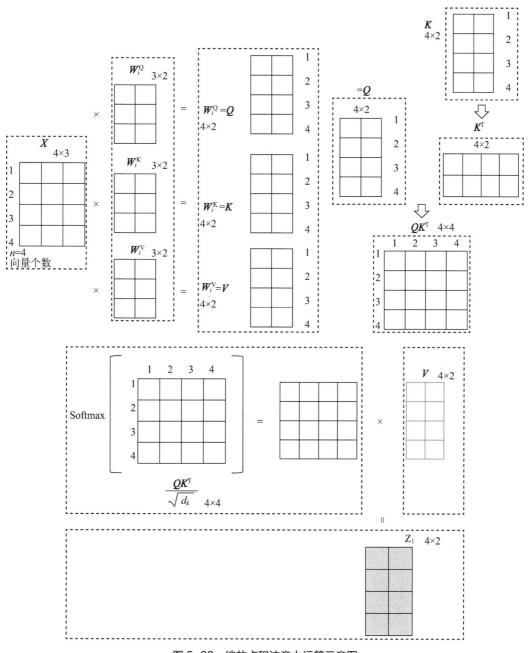

图 5-39　缩放点积注意力运算示意图

我们如果把缩放点积注意力机制并行处理 h 次（也就是 h 个注意力头）便是多头注意力机制，如图 5-40 所示，Q、K、V 分别乘以对应不同参数变换矩阵 W_q、W_k、W_v

（图中通过 Linear 层实现）得到多个输出，也就是拆分出多个头，再输入到自注意力模块中，然后将多个输出张量拼接（Concat）成为一个张量，最后乘以参数矩阵 W_O，调整维度后输出（图中通过 Linear 层实现）。

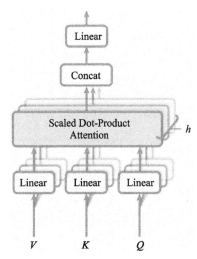

图 5-40　多头注意力机制运算示意图[①]

5.1.20　新算子和模型的探索

本章节讨论了多个算子和模型的原理和使用方法，它们都在不断创新和优化，并且衍生出新的算子和模型。比如开创性的 Transformer（式 5.1-1）。

$$\text{Attention}(Q, K, V) = \left[\text{Softmax}(QK^T / \sqrt{d_k})\right] V$$

这里的 Q、K、V 是怎么搞出来的？天才的思维啊！

这就好比杨过在吸取古墓派武功、九阴真经、独孤求败剑法的基础上自创了黯然销魂掌，和降龙十八掌有的一拼，成为我们的偶像——"神雕大侠"。

笔者认为人工智能领域工程师的成长包括以下几个阶段：

（1）入门阶段：能用目前开源的模型，进行微小改动后去训练我们的模型，并且满足我们的需求。

（2）提升阶段：用不同的算子和不同的网络模块去搭建一个新的模型，然后训练和评估它。

（3）专家阶段：设计新的算子和模型，把产品的效果和性能都发挥到了行业最佳的水平。

（4）行业引领者：开辟了一个新的领域，具备开创性创新。比如 YOLO、Transformer 等。

为了得到更好的模型效果和性能，探索新的算子和模型具有重要价值，但也充满挑战。我们需要多读些论文，不在多而在精（https://arxiv.org/、https://www.connectedpapers.com/）；

[①]　图来源：Vaswani A, Shazeer N, Parmar N, et al. Attention Is All You Need. arXiv, 2017.

关注领域的最新研究进展；多尝试新的想法（或者灵感）；研究重要的项目底层源码（https://paperswithcode.com/、Github）；不要止步于此，往前多迈一步，去思考和研究。此外需要培养自己的中文和英文读写能力（推荐一款语法纠正和校对工具 https://www.grammarly.com/）。

新算子和模型的探索，是艰苦的，也是甜蜜的，"路漫漫其修远兮，吾将上下而求索"。

5.2 剪枝

模型网络结构的剪枝，简称模型剪枝（Pruning），它是针对模型网络结构的精简，主要方法是去除冗余的连接或通道、神经元等。模型剪枝依赖处理器的特性，针对处理器的特性去剪枝才能提高性能。

根据剪枝的细粒度可将模型剪枝分为结构化剪枝（Structured Pruning）和非结构化剪枝（Unstructured Pruning）。

5.2.1 结构化剪枝

（1）滤波器剪枝（Filter Pruning），通过卷积核数量的减少来实现模型结构的简化，达到提高计算性能的目的。当然也会影响这个层的输出通道数，所以它的下一层也需要调整，如图 5-41 所示。

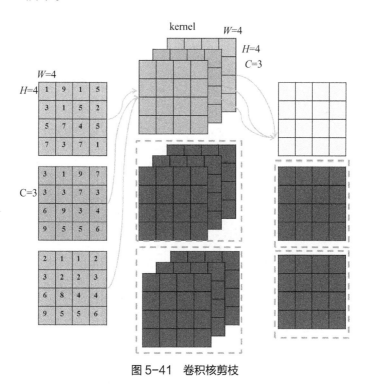

图 5-41 卷积核剪枝

（2）通道剪枝（Channel Pruning），通过卷积核的通道数的减少来简化模型结构，如图 5-42 所示。

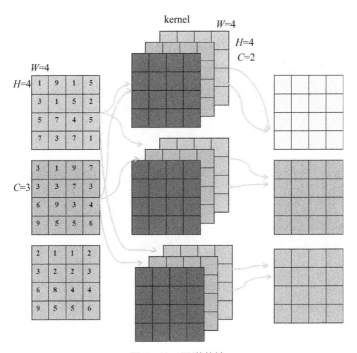

图 5-42 通道剪枝

（3）层剪枝（Layer Pruning），如图 5-43 所示，裁剪掉卷积层 conv0_2。

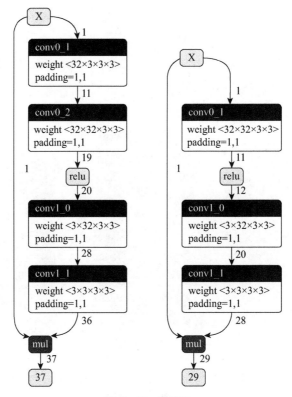

图 5-43 层剪枝

5.2.2 非结构化剪枝

（1）连接权重或神经元剪枝

将部分神经元之间的连接置零为连接权重剪枝，如图5-45中①处所示；将部分神经元的输出置零为神经元剪枝，如图5-44中②处所示。

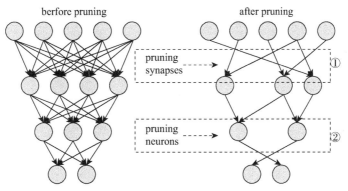

图5-44　连接权重或神经元剪枝[①]

图5-45对于我们理解卷积算子还不是很直观，我们可以进一步通过图5-45去理解。

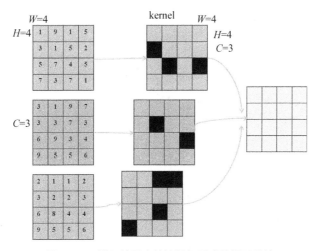

图5-45　卷积算子中的连接权重或神经元剪枝

（2）卷积核的部分通道内部分行或列向量剪枝，如图5-46所示。

---

① 图来源：Han S，Pool J，Tran J，et al. Learning both Weights and Connections for Efficient Neural Networks. MIT Press，2015.

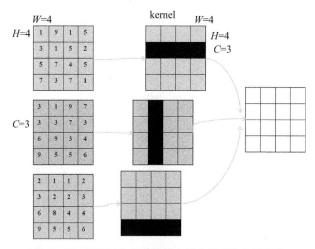

图 5-46 卷积核的部分通道内部分行或列向量剪枝

（3）卷积核某通道的剪枝，如图 5-47 所示。

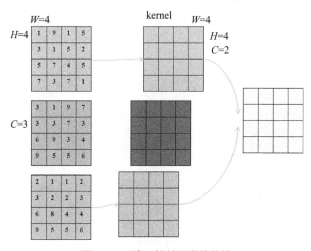

图 5-47 卷积核某通道的剪枝

非结构化剪枝会导致模型网络结构变得不规则，表现出结构稀疏化，通用硬件平台往往不一定支持加速。

5.3 网络架构搜索

前文中我们讲述的轻量化网络结构设计，一般采取人工设计，不仅需要依赖先验知识和经验，而且还需要不断地试错，这样对于先验借鉴少的新模型任务开发，往往试错成本很高，模型的开发效率很低。如何实现高效开发呢？NAS（Neural Network Architecture Search，神经网络架构搜索）实现代替人工设计网络结构的工作，用搜索算法自动寻找最合适的网络结构。NAS 是 AutoML 中重要的组成模块。AutoML（Automated

Machine Learning,自动化机器学习)实现 AI 模型自动化流水线训练开发,它包括训练参数(如学习率、调整学习率的策略、batch 大小、IOU、权值衰减等)、定义网络结构参数(网络层数、算子种类、卷积类型、卷积核大小等,一般用 NAS 实现)等。AutoML 框架有 TPOT、Auto-Sklearn、Hyperopt、H2O、AutoKeras、NNI、Vega[①] 等。

5.4 低秩分解

低秩分解(Low Rank Filters)也叫低秩近似,它实现将大权重矩阵分解成多个小矩阵,比如采用两个 $K \times 1$ 卷积核替换 $K \times K$ 卷积核,实现减少权值张量参数和去除冗余,其原理是卷积核具有低秩特性。低秩分解的方法主要有奇异值分解(Singular Value Decomposition,SVD)、Tucker 分解、CP 分解等,图 5-48 为卷积的低秩分解示意图。鉴于笔者目前在实际项目应用中涉及不多,故在本书中不做进一步讲述,读者朋友可以参考相关的资料做进一步的研究和实践。

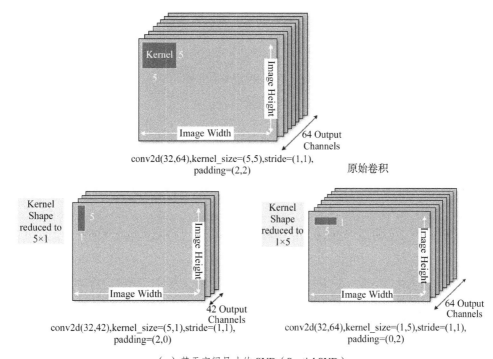

(a)基于空间尺寸的 SVD(Spatial SVD)

---

① 可以进一步参考以下资料:
网址:https://github.com/rhiever/tpot;https://github.com/automl/auto-sklearn;https://autokeras.com/;https://nni.readthedocs.io/zh/latest/nas.html;https://github.com/huawei-noah/vega
论文:[1] Yao Q, Wang M, Chen Y, et al. Taking Human out of Learning Applications: A Survey on Automated Machine Learning. 2018.
[2] Zller, Marc-André, Huber M F. Benchmark and Survey of Automated Machine Learning Frameworks. 2019.

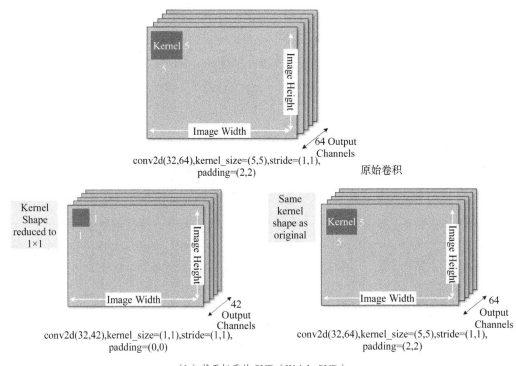

(b)基于权重的 SVD(Weight SVD)

图 5-48　卷积的低秩分解示意图[1]

5.5　知识蒸馏

由于移动端或嵌入式设备上 CPU 处理性能和存储空间的限制，部署的模型对体积要求不能太大，需要把模型做进一步瘦身。我们如何将瘦身后的模型尽可能接近瘦身前的模型呢？于是我们又有一个大招——知识蒸馏。

知识蒸馏（Knowledge Distillation，KD）也是一种模型压缩的方法，它能用于解决如何训练一个小模型（轻量化模型网络），并且取得很好的模型精度的问题。知识蒸馏的一种方法：让教师网络去指导学生网络在模型训练中学习，学习教师网络的知识［包含暗知识（dark knowledge）］。这里的学生网络（Student Model）称为小模型，它的模型复杂度和参数较小；教师网络（Teacher Model）称为大模型，往往模型网络结构比较复杂，模型参数较大。知识蒸馏带来的直接好处是：加速小模型的训练收敛速率。其他好处是通过知识蒸馏得到的小模型也接近大模型的泛化能力。如图 5-49 所示为一种知识蒸馏的训练方法，我们分别将教师网络和学生网络输出结果建立一个损失函数（记为Loss1），学生网络推理输出结果和 Ground truth 之间建立另一个损失函数（记为 Loss2），通过 Loss1 和 Loss2 一起去约束学生网络的训练。

---

[1]　图来源：AIMET 开源库

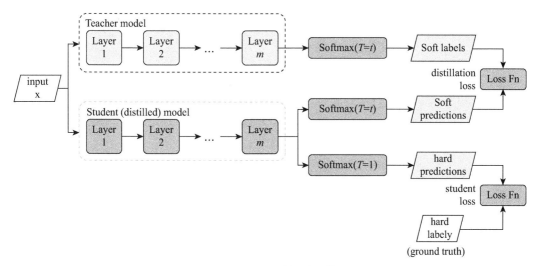

图 5-49 知识蒸馏流程①

知识蒸馏的知识除了上文基于教师网络的输出结果，即输出特征知识（Response-based Knowledge），还包括教师网络的中间特征知识（Feature-based Knowledge）和关系特征知识（Relation-based Knowledge），基于这些知识可以去建立适合的损失函数去训练小模型，如图 5-50 所示。

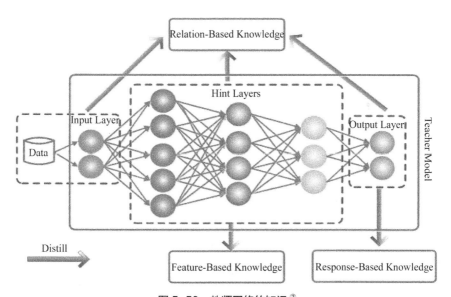

图 5-50 教师网络的知识②

知识蒸馏的方法有多种，除了上文介绍的教师 – 学生蒸馏，更多方法如图 5-51 所示。

---

① 图来源：https://intellabs.github.io/distiller/knowledge_distillation.html
② 图来源：Gou J, Yu B, Maybank S J, et al. Knowledge Distillation: A Survey. 2020.

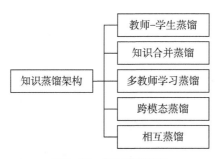

图 5-51 知识蒸馏架构

此外，下一节的量化中涉及的量化感知训练也属于知识蒸馏的一种方法，实现将高位宽的模型中的知识向低位宽的模型的转移。

5.6 量化

模型量化是将模型中用 float32（FP32）类型表示的浮点权重和激活参数用 float16（FP16）、int8（INT8）类型或 4bit（2bit、1bit）低比特数来表示（如表 5-2 所示）。量化是模型压缩的利器。

表 5-2 FP32、FP16、INT8 类型取值范围

数据类型	取值范围	最小正数值
FP32	$-3.4 \times 10^{38} \sim +3.4 \times 10^{38}$	1.4×10^{-45}
FP16	$-65504 \sim +65504$	5.96×10^{-8}
INT8	$-128 \sim +127$	1

量化的好处有以下几点：

（1）减小模型的体积，进而减少内存和存储空间占用。例如使用 float16 量化后，模型体积缩减一半；使用 INT8 量化后，模型体积缩减为原来的四分之一。

（2）减小计算量和访存量。一般整数运算指令延迟（Latency）小于对应的浮点运算指令。

（3）减小处理器功耗。一般处理器的功耗主要包括计算和访存两个部分。关于功耗问题，可以结合表 5-3 和图 5-52 讨论。

从表 5-3 可以看出同样是 32 bit 的 ADD 运算，float 型运算的功耗是 int 型运算的功耗的 9 倍，而 DRAM Memory 访存的功耗是 int 型 ADD 运算的 6400 倍，可见访存功耗是大头。

表 5-3 处理器实现加法、乘法等操作的功耗

Operation	Energy/pJ	Relative Cost
32 bit int ADD	0.1	1
32 bit float ADD	0.9	9
32 bit Register File	1	10
32 bit int MULT	3.1	31
32 bit float MULT	3.7	37
32 bit SRAM Cache	5	50
32 bit DRAM Memory	640	6400

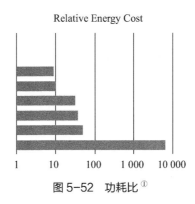

图 5-52 功耗比[①]

图 5-53（a）、(b)、(c) 表示 DRAM 访存功耗为 1.3～2.6 nJ（1 nJ = 1000 pJ），远高于整型运算、浮点运算功耗。图（a）中 8 bit 的 Add 的功耗为 0.03 pJ，而 32 bit 的 Add 的功耗为 0.1 pJ，可见低比特的运算功耗低于高比特的运算功耗。图（d）表示指令功耗分解（Instruction Energy Breakdown），其中 Add 运算功耗为 70 pJ，一级指令 Cache 访问功耗为 25 pJ，寄存器访问功耗为 6 pJ。

Integer		FP		Memory	
Add		FAdd		Cache	(64bit)
8 bit	0.03pJ	16 bit	0.4pJ	8KB	10pJ
32 bit	0.1pJ	32 bit	0.9pJ	32KB	20pJ
Mult		FMult		1MB	100pJ
8 bit	0.2pJ	16 bit	1.1pJ	DRAM	1.3-2.6pJ
32 bit	3.1pJ	32 bit	3.7pJ		

（a）整型运算功耗　　（b）浮点运算功耗　　（c）访存功耗

[①] 图来源：Han S, Pool J, Tran J, et al. Learning both Weights and Connections for Efficient Neural Networks. MIT Press, 2015.

（d）指令功耗分解

图 5-53　功耗对比[1]

量化是模型部署的重要优化工具。有读者朋友可能有这样的疑问：我们为什么不直接在模型训练环节中直接训练一个 FP16 或者 INT8 类型的模型呢？原因是在使用目前的深度学习框架训练模型中，其梯度值一般都很小，基本都是小数值，如果用 FP16、INT8 可能导致浮点精度不够，不能满足梯度表示和反向传播的要求，往往训练很慢、很难收敛，甚至不收敛。但是在推理中，由于只需要前向传播，如果在能够满足一定的精度要求下，就有可能实现用 FP16、INT8 等类型表示权重和偏置等参数。所以模型量化在模型的推理环节中是重要的优化方法，比如 TensorRT[2]、TNN、QNN/SNPE、TFLite 等推理框架中都得到应用以及 Qualcomm 推出的支持模型压缩和量化的工具 AIMET（AI Model Efficiency Toolkit）等等。

由于量化可能会导致精度的损失，且损失程度的敏感度和模型的任务种类有很大关联，一般在计算机视觉任务中分类优于目标检测，目标检测优于分割，也就是说分割任务模型对量化的精度损失更加敏感。

5.6.1　量化原理

量化的本质是实现两个数据集合的映射关系，也就是函数关系[3]。如果映射中的量化间隔采用相同的值，则为均匀量化（uniform quantization），也叫线性量化，反之为非均匀量化（non-uniform quantization），比如对数量化。

一般在实际工程中使用均匀量化，如图 5-54 所示，根据线性等比例性可以得到以下公式：

$$\frac{Q_{max} - Q_{min}}{V_{max} - V_{min}} = \frac{Q_{max} - x}{V_{max} - x} \quad (5.6\text{-}1)$$

---

[1] 图来源：Horowitz M.1.1 Computing's energy problem（and what we can do about it）. 2014 IEEE International Solid-State Circuits Conference（ISSCC）. IEEE，2014.

[2] 用 KL 散度（Kullback-Leibler Divergence，KLD）确定 |T| 实现饱和（Saturate）量化，进一步可参考 8-bit Inference with TensorRT，https://on-demand.gputechconf.com/gtc/2017/presentation/s7310-8-bit-inference-with-tensorrt.pdf

[3] 对于 A 和 B 两个非空集合，如存在一个法则 f（），使 A 中每个元素 x 在 B 中有唯一确定的元素 y 与之对应，则称 f 为从 A 到 B 的映射，用函数表示为 f（）：$A \rightarrow B$。

其中 $V_{\min}$、$V_{\max}$ 分别为待量化参数的最小值和最大值，$Q_{\min}$、$Q_{\max}$ 分别为量化后的最小值和最大值，x 为待量化参数，x_q 为量化后的参数

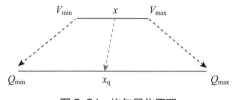

图 5-54　均匀量化原理

我们把 $\dfrac{V_{\max}-V_{\min}}{Q_{\max}-Q_{\min}}$ 定义为 scale，代表量化后的定点数值可表示的最小分度值，也就是上文讲到的量化间隔。接下来式（5.6-1）可进一步变换如下：

$$\dfrac{1}{\text{scale}} = \dfrac{Q_{\max} - x_q}{V_{\max} - x}$$

$$\dfrac{V_{\max} - x}{\text{scale}} = Q_{\max} - x_q$$

$$x_q = Q_{\max} - \dfrac{V_{\max} - x}{\text{scale}}$$

$$x_q = \dfrac{x}{\text{scale}} + (Q_{\max} - \dfrac{V_{\max}}{\text{scale}}) = \dfrac{Q_{\max}(x - V_{\min}) - Q_{\min}(x - V_{\max})}{V_{\max} - V_{\min}} \quad (5.6\text{-}2)$$

最终得到量化公式如下：

$$x_q = \text{round}\left(\dfrac{x}{\text{scale}} + z\right) \quad (5.6\text{-}3)$$

其中 round 实现取整操作，z 值为 $z = Q_{\max} - \dfrac{V_{\max}}{\text{scale}}$。

反量化公式如下：

$$x = (x_q - z) \cdot \text{scale} \quad (5.6\text{-}4)$$

当 $Q_{\min} = 0$，$Q_{\max} = 255$，也就是 UINT8 量化，得到量化公式如下：

$$x_q = \text{round}\left(\dfrac{x}{\text{scale}} + z\right),\ z = 255 - \dfrac{V_{\max}}{\text{scale}}$$

下面我们结合具体的待量化参数来验证以上的公式。已知待量化参数为 [-2.1，-0.9，0，1，2.5]，那么其中 $V_{\min} = -2.1$，$V_{\max} = 2.5$，$V_{\max} - V_{\min} = 2.5 - (-2.1) = 4.6$，则 scale = 4.6/（255-0）= 0.018039，z = 255 - 2.5/0.018039 = 116.41。

通过公式 $x_q = \text{round}\left(\dfrac{x}{\text{scale}} + z\right)$ 计算得到这组参数量化后的值为 [0，67，116，172，255]。

再通过反量化公式：

$$x = (x_q - z) \cdot \text{scale}$$

$$= x_q \cdot \text{scale} - z \cdot \text{scale}$$

$$= x_q \cdot \text{scale} - (Q_{max} - \frac{V_{max}}{\text{scale}}) \cdot \text{scale}$$

$$= x_q \cdot \text{scale} - (Q_{max} \cdot \frac{V_{max} - V_{min}}{Q_{max} - Q_{min}} - \frac{V_{max}}{\text{scale}} \cdot \text{scale})$$

$$= x_q \cdot \text{scale} - (255 \cdot \frac{V_{max} - V_{min}}{255} - V_{max})$$

$$= x_q \cdot \text{scale} + V_{min}$$

$$= x_q \cdot 0.018039 - 2.1$$

得到［0，67，116，172，255］反量化后的值为［-2.1，-0.891387，-0.007476，1.002708，2.499945］，对比之前的浮点数值［-2.1，-0.9，0，1，2.5］，可以看到有一点的精度偏差。

我们还可以通过高通 SNPE 的量化来验证一下量化的计算方法。如图 5-56 所示，我们选择其中的卷积算子来分析。从图中我们看到 min = -5.015156745911，max = 3.078925371170，scale = 0.031741499901，offset = -158.000000000000，其中的 offset 也就是上文中的 z 偏移量。接下来我们用上文的公式验证一下，验证过程如下：

$$\text{scale} = \frac{V_{max} - V_{min}}{Q_{max} - Q_{min}}$$

$$= \frac{3.078925371170 - (-5.015156745911)}{255 - 0}$$

$$= 0.03174149849835686$$

$$z = Q_{max} - \frac{V_{max}}{\text{scale}}$$

$$= 255 - 97.00000053020132$$

$$= 157.9999994697987$$

验证计算得到 scale 和 offset 值和图 5-55 的值一致。

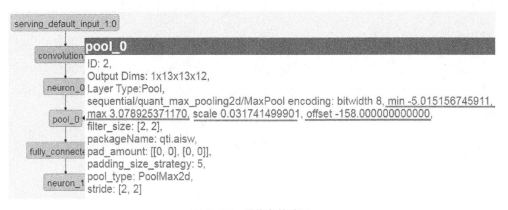

图 5-55 量化参数验证

5.6.2 对称量化 / 非对称量化

在上文的量化公式 $x_q = \text{round}(\frac{x}{\text{scale}} + z)$ 中：

当 $x = 0$，$x_q = 0$ 时为对称量化（Symmetric）；

当 $x = 0$，$x_q \neq 0$ 时为非对称量化（Asymmetric）。

模型的 Weights（权重）参数的数据分布多以 0 为中心对称，在量化中多采用对称量化，而 Activation（激活）一般采用非对称量化。

5.6.3 伪量化节点

前文中我们讨论了量化原理和对称量化 / 非对称量化，我们已经知道量化是通过量化参数 scale 和 zero point 来进行量化映射计算，这个量化参数是通过需要量化的数据的最小值和最大值（min-max）来计算的。

模型网络中的节点包括输入节点和算子（输入节点可以看作特殊的算子），其中某一个算子，如果需要量化 Weights，则需要添加一个伪量化节点（fake quantization node），用于保存 Weights 参数的 min-max（为什么叫保存，因为这个参数一般在模型量化环节就计算好，而不是在推理中再计算），进而已经量化好的模型在推理中通过量化位宽和量化方法进行量化映射计算。如果某一个算子，需要 Activation（激活）量化，则需要添加一个用于保存算子输出的 min-max 的伪量化节点，用于实现模型推理中某一个算子的输出量化。

5.6.4 训练后量化 / 量化感知训练

量化按照是否需要继续训练可分为训练后量化（Post-Training Quantization，PTQ）[1] 和量化感知训练（Quantization-Aware Training，QAT）。

训练后量化是直接对已经训练好的模型进行量化，一般只需要少量的数据用于量化参数的计算。这种量化方式可能会造成量化精度的丢失。图 5-56 为训练后量化流程。

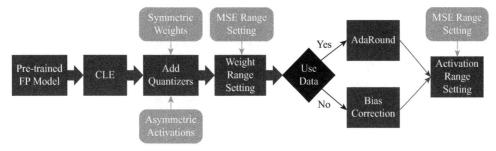

图 5-56 训练后量化流程[2]

---

[1] 更多资料可以参考 https://tensorflow.google.cn/lite/performance/hexagon_delegate?hl=zh-cn

[2] 图来源：Nagel M, Fournarakis M, Amjad R A, et al.A White Paper on Neural Network Quantization. 2021. DOI:10.48550/arXiv.2106.08295.

量化感知训练是在已经训练好的模型中添加伪量化节点，然后通过一定量的数据用于模型的 finetune（微调）训练，微调模型的权重参数、偏置参数以及量化参数（min-max）。通过量化感知训练可以很好地解决模型量化造成的精度丢失问题。量化感知训练对学习率比较敏感，设置过大，容易导致模型退化。笔者测试中发现，如果将学习率设置为 0.001，则模型直接退化；学习率设置为 0.0001，模型可以收敛；学习率为 0.000001，感知训练中模型收敛很好。图 5-57 为量化感知训练流程。

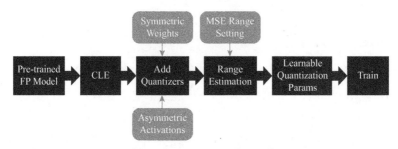

图 5-57　量化感知训练流程[①]

5.6.5　量化提升策略

（1）按张量量化 / 按通道量化

按张量量化（Per-Tensor Quantization）是将模型中 Weight（权重）或 Activation 对应的每一个 Tensor 整体量化，每个 Tensor 对应一组量化参数 min/max（通过 min/max 可以得到量化映射参数：比例因子 S 和零点参数 zero-point）。

按通道量化（Per-Channel Quantization）是将模型中 Weight 或 Activation 对应的每一个 Tensor 按照其通道分别进行量化，于是每个 Tensor 的 n 个通道对应的量化参数 min/max 有 n 个组。比如卷积核 Weight 如果按通道量化，那么这个卷积核的每个通道将对应一组量化参数。

一般 Weight 采用对称按通道量化，Activation 采用非对称按张量量化。

（2）跨层均衡化

讨论跨层均衡化之前，我们先介绍箱线图（Box plot），显而易见，得名于图中的小矩阵块，即箱子，如图 5-58 所示，它用于反映数据的中心位置和分布情况。

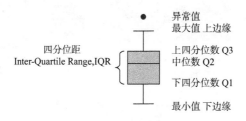

图 5-58　箱线图

---

① 图来源：Nagel M, Fournarakis M, Amjad R A, et al.A White Paper on Neural Network Quantization. 2021. DOI:10.48550/arXiv.2106.08295.

我们把一组数据从小到大排列后，其中前 75% 的数据对应箱线图的 Q3 位置（箱子的上底），前 50% 的数据对应箱线图的 Q2 位置（也叫中位数，它是位于箱子的中间的一条线，代表样本数据的平均水平），前 25% 的数据对应箱线图的 Q1 位置（箱子的下底），这组数据的最大值（max）和最小值（min）分别对应箱线图的上下边缘（此处不考虑异常值）。所以箱子里（Q3 到 Q1 之间）包含 50% 的数据。

图 5-59（a）为 MobileNet V2 模型网络中一个可分离卷积层输出的 32 个卷积核的权重参数箱形图，包含 max、min、Q3、Q1 及中位数信息，可以看出每个通道权重范围差异很大。如果采用按张量量化，得到的量化参数 max/min 对于其中一些分布范围很小的通道，量化后可能压缩为几个值，这样量化误差会很大。而采用按通道量化，需要为每个通道分别计算一组量化参数 max/min，虽然可以解决每个通道权重分布差异的问题，但是有时候对计算性能开销很大。

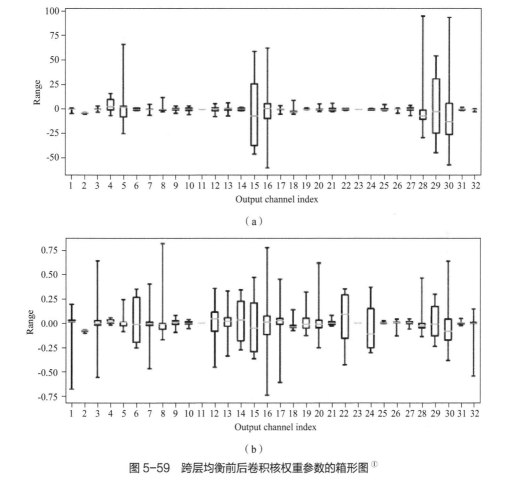

图 5-59 跨层均衡前后卷积核权重参数的箱形图[①]

---

① 图来源：https://quic.github.io/aimet-pages/releases/latest/user_guide/post_training_quant_techniques.html#ug-post-training-quantization

采用跨层均衡（Cross-Layer Equalization，CLE）的目的就是实现调整通道的权重参数的分布范围（其原理可以参考论文 *Data-Free Quantization Through Weight Equalization and Bias Correction*），图 5-59（b）为均衡后输出的 32 个通道权重参数的箱形图，可以看出大多数通道的权重参数分布范围从原先的 −25～95 调整到 −0.7～0.8，每个卷积核的权重参数都在差不多的范围内。

（3）其他量化提升策略简述

量化提升策略还有 AdaRound[①]/Nearest Rounding、Bias Correction、量化位宽、量化误差分析等。这里以 AIMET 量化方法为例简单讨论量化位宽。量化位宽的测试代码如下：

```
from aimet_common.defs import QuantScheme
from aimet_tensorflow.quantsim import QuantizationSimModel
sim = QuantizationSimModel（model = net,
            quant_scheme = QuantScheme.post_training_tf,
            rounding_mode = " nearest",
            default_output_bw = 8,
            default_param_bw = 8）
```

以上代码将 Weights 量化位宽设置为 INT8，Activation 量化位宽设置为 INT8，这种模式叫作 W8A8 量化。

然后将测试代码修改如下：

```
sim = QuantizationSimModel（model = net,
            quant_scheme = QuantScheme.post_training_tf,
            rounding_mode = " nearest",
            default_output_bw = 16,
            default_param_bw = 8）
```

以上代码将 Weights 量化位宽设置为 INT8，Activation 量化位宽设置为 INT16，这种模式叫作 W8A16 量化。

当然 W4A4 量化模式也是有的。选择这种模式，需要做到硬件、性能和精度的均衡。

5.6.6 量化感知训练框架介绍

（1）TensorFlow 量化感知训练

剪枝、量化感知训练和知识蒸馏训练，是轻量化模型部署三板斧。TensorFlow 提供一套模型优化的方案，支持模型剪枝和量化。TensorFlow 框架中通过 tensorflow-model-

---

① 更多资料可以参考 https://quic.github.io/aimet-pages/releases/latest/user_guide/adaround.html#ug-adaround optimization 实现量化感知训练[1]。

第一步，安装 pip install tensorflow-model-optimization。

第二步，使用 tfmot.quantization.keras.quantize_model 给模型添加 FakeQuant（伪量化）节点，用于下一步的量化感知训练，其关键代码如下。

```
import tensorflow_model_optimization as tfmot  # 导入量化工具包
fp32_model = tf.keras.models.load_model（'fp32_model.h5'）
quant_aware_model = tfmot.quantization.keras.quantize_model（fp32_model）
quant_aware_model.summary（）
```

如果有些层量化后精度损失很大，可以采用选择性量化模型的某些层的方案，目的是实现模型准确率、推理时延和模型大小的均衡。关键代码如下。

```
quantize_annotate_layer = tfmot.quantization.keras.quantize_annotate_layer
input = tf.keras.Input（shape=（10,））
x = quantize_annotate_layer（tf.keras.layers.Dense（2））（input）
output = tf.keras.layers.Flatten（）（x）
model = tf.keras.Model（inputs=input, outputs=output）
quant_aware_model = tfmot.quantization.keras.quantize_apply（model）
quant_aware_model.summary（）
```

目前 tfmot.quantization.keras.quantize_annotate_layer 添加伪节点只支持 tf.keras.layers 中的方法，所以我们的模型尽量用 tf.keras.layers 方法。例如我们用 layers.AveragePooling2D 替换 tf.nn.avg_pool，用 layers.Multiply（）替换"×"，用 tf.keras.layers.Add 替换"+"，用 tf.keras.layers.Permute 替换 tf.transpose，等等，具体代码如下所示[2]。

```
import tensorflow as tf
from tensorflow.keras.layers import AveragePooling2D, Multiply, Add, Permute
x = tf.random.normal（（1, 640, 320, 3））
y1 = tf.nn.avg_pool（x, ksize=[1, 2, 2, 1], strides=[1, 2, 2, 1], padding='VALID'）
```

---

[1] TensorFlow 官网和代码仓库：https://tensorflow.google.cn/model_optimization/guide/quantization/training_comprehensive_guide?hl=zh_cn，

https://github.com/tensorflow/model-optimization/blob/master/tensorflow_model_optimization/g3doc/guide/quantization/training_example.ipynb

其中 .ipynb 文件可以使用命令 jupyter nbconvert--to script training_comprehensive_guide.ipynb 转换为 Python 文件
https://tensorflow.google.cn/model_optimization/guide/quantization/training_comprehensive_guide?hl=zh-cn
https://tensorflow.google.cn/model_optimization/api_docs/python/tfmot/all_symbols
https://tensorflow.google.cn/model_optimization/guide/quantization/training?hl=zh-cn

[2] 更多使用可以参考 https://tensorflow.google.cn/api_docs/python/tf/keras/layers

```
y2 = AveragePooling2D（pool_size =（2，2），strides =（2，2），padding ='VALID'，
data_format = None )( x )
# pool_size：池化窗口大小
# strides：池化窗口在每一维度上移动步长，默认等于 pool_size
# padding：'VALID'无填充，'SAME'用 0 填充
# data_format：输入张量维度顺序，默认为［batch，height，width，channel］
y1==y2

x1 = tf.random.normal（（1，640，320，3））
x2 = tf.random.normal（（1，1，1，3））
y1 = x1*x2
y2 = tf.keras.layers.Multiply（ ）([x1，x2］）
y1==y2

x1 = tf.random.normal（（1，640，320，3））
x2 = tf.random.normal（（1，640，320，3））
y1 = x1+x2
y2 = tf.keras.layers.Add（ ）([x1，x2］）
y1==y2

x = tf.random.normal（（1，640，320，3））
y1 = tf.transpose（x，perm=［0，1，3，2］）
y2 = tf.keras.layers.Permute（dims=（1，3，2））( x )
y1==y2
```

（2）AIMET 量化感知训练

基于高通平台可以使用 AIMET 实现量化感知训练，如图 5-60 所示为 AIMET 量化流程。

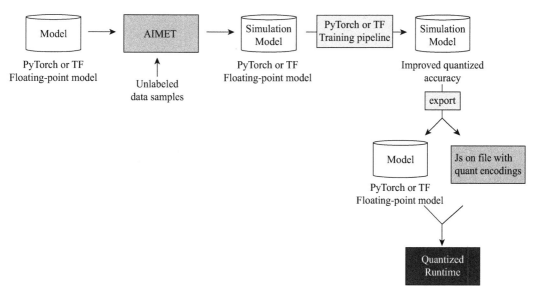

图 5-60　AIMET 量化流程[1]

AIMET 感知训练准备工作：

第一步，创建虚拟环境，方法如下。

(base)user@user:~$ conda create -n aimet python = 3.8.13
\# To activate this environment, use
\#
\# $ conda activate aimet
\#
\# To deactivate an active environment, use
\#
\# $ conda deactivate
aimet$ python
Python 3.8.17 | packaged by conda-forge |(default, Jun 16 2023, 07:06:00)
[GCC 11.4.0] on linux
Type" help", " copyright", " credits" or" license" for more information.
\>\>\>

第二步，下载 AIMET 安装包和安装，如图 5-61 所示，下载地址为 https://github.com/quic/aimet/releases。

---

[1] 图来源：https://quic.github.io/aimet-pages/releases/1.26.0/user_guide/model_quantization.html#aimet-quantization-workflow

aimet_onnx-onnx_cpu_1.31.0-cp38-cp38-linux_x86_64.whl

aimet_onnx-onnx_gpu_1.31.0-cp38-cp38-linux_x86_64.whl

aimet_tensorflow-tf_cpu_1.31.0-cp38-cp38-linux_x86_64.whl

aimet_tensorflow-tf_gpu_1.31.0-cp38-cp38-linux_x86_64.whl

aimet_torch-onnx_cpu_1.31.0-cp38-cp38-linux_x86_64.whl

aimet_torch-onnx_gpu_1.31.0-cp38-cp38-linux_x86_64.whl

aimet_torch-torch_cpu_1.31.0-cp38-cp38-linux_x86_64.whl

aimet_torch-torch_cpu_pt19_1.31.0-cp38-cp38-linux_x86_64.whl

aimet_torch-torch_gpu_1.31.0-cp38-cp38-linux_x86_64.whl

LICENSE.pdf

Source code (zip)

Source code (tar.gz)

图 5-61　AIMET 安装包

安装命令如下：

sudo apt-get install liblapacke liblapacke-dev libjpeg8-dev zlib1g-dev

$ pip install torch==1.9.1 torchvision==0.10.1 -i https://pypi.tuna.tsinghua.edu.cn/simple/

$ pip install AimetCommon-torch_cpu_1.31.0-cp38-cp38-linux_x86_64.whl -i https://pypi.tuna.tsinghua.edu.cn/simple/

$ pip install AimetTorch-torch_cpu_1.31.0-cp38-cp38-linux_x86_64.whl -f https://download.pytorch.org/whl/torch_stable.html

$ pip install Aimet-torch_cpu_1.31.0-cp38-cp38-linux_x86_64.whl -i https://pypi.tuna.tsinghua.edu.cn/simple/

$ sudo ln -s /usr/lib/x86_64-linux-gnu/libjpeg.so /usr/lib

第三步，克隆 AIMET 工程代码和修改量化感知训练代码。运行命令 git clone https://github.com/quic/aimet.git，里面有量化感知训练的示例代码。在路径 aimet/Examples/torch/quantization 中我们运行命令 jupyter nbconvert--to script qat.ipynb 实现 qat.ipynb 文件转化为 qat.py 文件，参考这个代码可以很快上手使用 AIMET 实现量化感知训练。AIMET 量化感知训练代码如下。

```
from aimet_torch.model_preparer import prepare_model
model = prepare_model（model）

from aimet_torch.batch_norm_fold import fold_all_batch_norms
_ = fold_all_batch_norms（model, input_shapes =（1, 1, 28, 28））

from aimet_common.defs import QuantScheme
from aimet_torch.quantsim import QuantizationSimModel
```

```
sim = QuantizationSimModel (model = model,
                           quant_scheme = QuantScheme.post_training_tf_enhanced,
                           dummy_input = dummy_input,
                           default_output_bw = 8,
                           default_param_bw = 8)
sim.compute_encodings (forward_pass_callback = pass_calibration_data,
                       forward_pass_callback_args = use_cuda)
sim.model.train ()
# 模型训练
dummy_input = dummy_input.cpu ()
sim.export (path = './output/', filename_prefix = 'model_mnist_after_qat',
dummy_input = dummy_input)
```

AIMET 量化感知训练后得到的相关文件，如图 5-62 所示。

图 5-62　AIMET 量化感知训练后得到的相关文件

其中文件 model_mnist_after_qat.onnx 的网络结构如图 5-63 所示，包括 conv1、relu1、conv2、module_avg_pool2d 和 module_avg_pool2d#1.end：

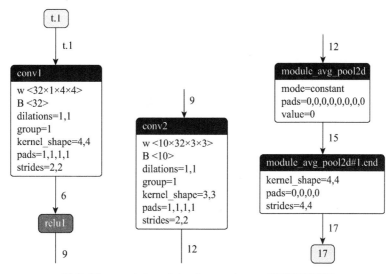

图 5-63　model_mnist_after_qat.onnx 的网络结构图

其中文件 model_mnist_after_qat_torch.encodings 包括两个部分，分别为运算符输出的量化参数"activation_encodings"和权重、偏差 Tensor 的量化参数"param_encodings"，具体如下。

```
{
  "activation_encodings" : {
    "conv1" : {
      "input" : {
        "0" : {
          "bitwidth" : 8,
          "dtype" : "int",
          "is_symmetric" : "False",
          "max" : 0.9941482543945312,
          "min" : 0.0,
          "offset" : 0,
          "scale" : 0.00389862060546875
        }
      }
    },
    "conv2" : {
      "output" : {
        "0" : {
          "bitwidth" : 8,
          "dtype" : "int",
          "is_symmetric" : "False",
          "max" : 0.48209184408187866,
          "min" : -0.5013754963874817,
          "offset" : -130,
          "scale" : 0.0038567346055060625
        }
      }
    },
    "module_avg_pool2d" : {
      "output" : {
        "0" : {
```

```
            "bitwidth": 8,
            "dtype": "int",
            "is_symmetric": "False",
            "max": 0.16692544519901276,
            "min": -0.1877911239862442,
            "offset": -135,
            "scale": 0.001391045399941504
          }
        }
      },
      "relu1": {
        "output": {
          "0": {
            "bitwidth": 8,
            "dtype": "int",
            "is_symmetric": "False",
            "max": 1.443389654159546,
            "min": 0.0,
            "offset": 0,
            "scale": 0.005660351365804672
          }
        }
      }
    },
    "excluded_layers": [],
    "param_encodings": {
      "conv1.weight": [
        {
          "bitwidth": 8,
          "dtype": "int",
          "is_symmetric": "True",
          "max": 0.24986734986305237,
          "min": -0.25183477997779846,
          "offset": -128,
```

```
          "scale" : 0.0019674592185765505
        }
      ],
      "conv2.weight" : [
        {
          "bitwidth" : 8,
          "dtype" : "int",
          "is_symmetric" : "True",
          "max" : 0.069852016866620712,
          "min" : -0.070402033362703323,
          "offset" : -128,
          "scale" : 0.0005500158877111971
        }
      ]
  },
  "quantizer_args" : {
    "activation_bitwidth" : 8,
    "dtype" : "int",
    "is_symmetric" : true,
    "param_bitwidth" : 8,
    "per_channel_quantization" : false,
    "quant_scheme" : "post_training_tf_enhanced"
  },
  "version" : "0.6.1"
}
```

TensorFlow 环境，AIMET 安装方法如下：

```
$ pip install tensorflow==2.4.3 keras==2.3.1 -i https://pypi.tuna.tsinghua.edu.cn/simple/
$ pip install AimetCommon-tf_cpu_1.31.0-cp38-cp38-linux_x86_64.whl
AimetTensorflow-tf_cpu_1.31.0-cp38-cp38-linux_x86_64.whl
Aimet-tf_cpu_1.31.0-cp38-cp38-linux_x86_64.whl -i
https://pypi.tuna.tsinghua.edu.cn/simple/
```

使用 AIMET 碰到的问题和解决方法：

问题 1：

ImportError: /lib/x86_64-linux-gnu/libm.so.6: version `GLIBC_2.29' not found
（required by /home/***/anaconda3/envs/***/lib/python3.8/site-packages
/aimet_common/libpymo.cpython-38-x86_64-linux-gnu.so）

解决方法：

在网址 http://ftp.gnu.org/gnu/glibc/glibc-2.29.tar.gz 下载 glibc-2.29.tar.gz 并解压 tar -zxvf glibc-2.29.tar.gz，安装 sudo apt-get install gawk。然后运行命令 mkdir build && cd build ./../configure --prefix=/usr/local/glibc，可能出现以下问题：

configure: error:
*** These critical programs are missing or too old: bison
*** Check the INSTALL file for required versions.

解决方法：

sudo apt install bison。

继续编译、安装和配置：make -j8 && sudo make install && cd /lib/x86_64-linux-gnu && sudo ln -sf /usr/local/glibc/lib/libm-2.29.so libm.so.6，验证问题得以解决。

问题 2：

tensorflow.python.framework.errors_impl.NotFoundError: /usr/lib/x86_64-linux-gnu/libstdc++.so.6: version `GLIBCXX_3.4.26' not found（required by /home/***/anaconda3/envs/***/lib/python3.8/site-packages/aimet_common/libaimet_tf_ops.so）

解决方法：

如图 5-64 所示运行命令 strings /usr/lib/x86_64-linux-gnu/libstdc++.so.6 | grep GLIBCXX，确定缺少 GLIBCXX_3.4.26 文件。

图 5-64　命令 strings /usr/lib/x86_64-linux-gnu/libstdc++.so.6 | grep GLIBCXX

在网址 http://www.vuln.cn/wp-content/uploads/2019/08/libstdc.so_.6.0.26.zip 下载 libstdc++.so.6.0.26 文件，按照以下命令配置，验证问题得以解决。

sudo mv libstdc++.so.6.0.26 /usr/lib/x86_64-linux-gnu/
cd /usr/lib/x86_64-linux-gnu/
sudo rm libstdc++.so.6 # 删除之前软连接
sudo ln libstdc++.so.6.0.26 libstdc++.so.6

第 6 章

端侧模型部署框架

模型部署是 AI 落地的重要环节。模型部署的目标设备有云端服务器、边缘设备、端侧设备，比如 PC、智能手机、可穿戴设备、车载、仪器仪表及其他嵌入式设备等。产品的需求，会决定模型部署方案的差异性。比如我们的模型对用户的隐私保护需求很大，那么端侧的部署就显得非常必要；又比如这个模型的确很小，或者终端设备的算力满足需求，结合功耗和产品需求评估，就可以考虑将模型部署在终端，进而可以降低开发和设备成本。模型的部署不应是单一的设备，特别是大模型时代，更是端、边、云的混合部署——协同工作，实现模型在精度、时延、功耗及隐私保护等维度的均衡。图 6-1 所示为混合 AI 架构，实现将语音采集的去噪和增强算法部署在终端设备中。

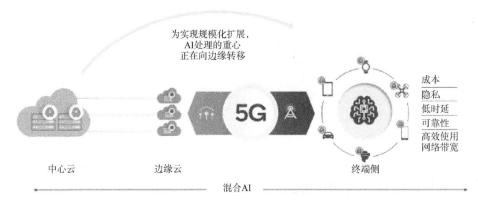

图 6-1　混合 AI [①]

本章主要介绍端侧模型部署框架，对于模型在端侧部署具有方向性指导，当然这些方案也不是局限在端侧，它们在云端和边缘端也一样具有工程可用性。

6.1　ARM ComputeLibray/Arm NN

ARM ComputeLibray（ACL）是基于 ARM 底层优化的机器学习 C++ 算子库，用于支持 Arm® Cortex®-A、Arm® Neoverse® 和 Arm® Mali™ GPUs 硬件架构。[②]

Arm NN 是 Arm Cortex-A CPUs 和 Arm Mali GPUs 上基于 Android 和 Linux 系统中开源的高性能的机器学习推理引擎。如图 6-2 所示，针对 Cortex-A CPU 和 Mali GPU 硬件内核，Arm NN 底层调用 ACL 库；针对 Arm Ethos-N NPU 硬件内核，Arm NN 底层调用 Ethos-N NPU 驱动程序；针对 Arm Cortex-M 处理器，Arm NN 底层调用 CMSIS-NN。[③]

---

① 图来源：混合 AI 是 AI 的未来，https://www.qualcomm.cn/。
② 更多资料可参考 https://github.com/ARM-software/ComputeLibrary。
③ 更多资料可参考 https://www.arm.com/products/silicon-ip-cpu/ethos/arm-nn；https://github.com/ARM-software/armnn.git。

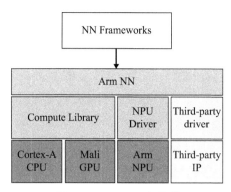

图 6-2 Arm NN 架构[①]

6.2 NNAPI

NNAPI（Android Neural Networks API）是 Android 平台的构建和运行神经网络的 API 接口，其系统架构如图 6-3 所示[②]：

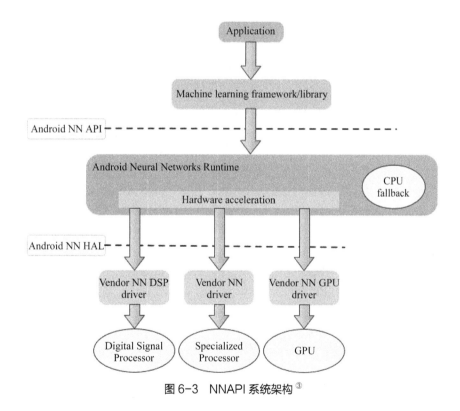

图 6-3 NNAPI 系统架构[③]

---

① 图来源：https://www.mlplatform.org/
② 更多资料可参考：https://developer.android.google.cn/ndk/guides/neuralnetworks/?hl=zh-cn，
https://source.android.google.cn/docs/core/architecture/modular-system/nnapi?hl=zh-cn
③ 图来源：https://developer.android.google.cn/ndk/guides/neuralnetworks/?hl=zh-cn

6.3 TensorRT

TensorRT 是 NVIDIA 推出的 C++ 语言开发的高性能的深度学习推理优化器，实现部署在 NVIDIA 显卡模型推理的最大吞吐量和推理效率。TensorRT 支持 TensorFlow、PyTorch、Caffe、MXNet 等深度学习框架。TensorRT 加速策略主要有模型参数量化（INT8、FP16 精度）、网络层间融合或张量融合和内核调整等。TensorRT-LLM 加速用于生成 AI 的最新大型语言模型（LLM），性能提高 8 倍，TCO（Total Cost of Ownership，总拥有成本）降低 5.3 倍，能耗降低近 6 倍[1]。

6.4 TNN/NCNN

TNN 是 Tencent 开源的深度学习移动端推理框架，支持 CPU（Armv7、Armv8、X86）、GPU（Mali、Adreno、Apple、NV GPU）、NPU 等硬件平台模型的部署，其推理架构如图 6-4 所示。

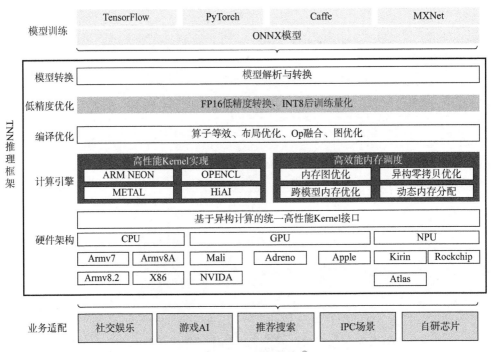

图 6-4　TNN 推理架构[2]

NCNN 是 Tencent 优图实验室开源的移动端深度学习推理框架，其特点是无第三方依赖、跨平台和支持大部分常用的 CNN 网络。

---

[1] 更多资料可以参考学习：https://developer.nvidia.com/zh-cn/tensorrt
[2] 图来源：https://github.com/Tencent/TNN.git

6.5 OpenVINO

OpenVINO 是针对 Intel 的硬件部署工具[1]。OpenVINO 支持 TensorFlow、PyTorch、ONNX 等深度学习框架模型，经过 OpenVINO 模型优化器优化适配后可以部署到 CPU、GPU、iGPU、VPU 平台，它的框架如图 6-5 所示。

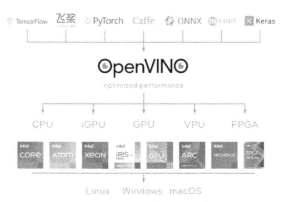

图 6-5 OpenVINO 框架[2]

6.6 TFLite

TFLite（TensorFlow Lite）是谷歌开源的轻量级深度学习推理框架，实现 TensorFlow 模型部署在移动端设备和嵌入式设备。TFLite 架构如图 6-6 所示。[3]

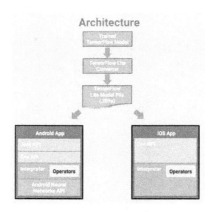

图 6-6 TFLite 架构[4]

---

[1] 更多资料可以参考：https://openvino.org/，https://docs.openvino.ai/latest/openvino_docs_OV_UG_Working_with_devices.html，https://docs.openvinotoolkit.org/latest/omz_models_model_handwritten_simplified_chinese_recognition_0001.html

[2] 图来源：https://docs.openvino.ai/latest/index.html

[3] TensorFlow 中模型文件格式包括：.pb（protocol buffer）文件：TensorFlow 保存的二进制模型文件，一般保存参数和网络结构，以便推理部署中无需定义网络结构；.ckpt 文件：TensorFlow 框架保存的模型文件；.h5（或 .hdf5）文件：tf.keras 保存的模型文件。

[4] 图来源：https://developers.googleblog.com/2017/11/announcing-tensorflow-lite.html

6.7　Core ML

Core ML 是 Apple 公司推出的基于 iOS 设备的移动端机器学习框架，它最大化优化内存占用和功耗。Core ML 首先需要将训练好的模型转化为 Core ML model，它的架构如图 6-7 所示。

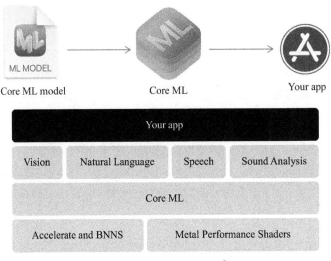

图 6-7　Core ML 架构[①]

6.8　RKNN SDK

RKNN SDK 为 Rockchip 提供的基于 NPU 平台的编程接口，用于部署和加速使用 RKNN-Toolkit 导出的 RKNN 模型。本书第 10 章，将以 NPU 加速 YOLOv8 模型推理为实例做进一步的讨论。

6.9　SNPE/QNN

SNPE（Snapdragon Neural Processing Engine SDK）是 Qualcomm 推出的神经网络处理引擎软件开发工具包，用于将网络模型部署在 Snapdragon 移动平台中的 Snapdragon CPU、Adreno GPU、Hexagon DSP 核中。SNPE 提供将 TensorFlow、PyTorch、ONNX 等模型转换为可加载到 SNPE 运行的 DLC 文件工具，具体流程如图 6-8 所示。[②]

---

① 图来源：https://developer.apple.com/documentation/coreml
② 更多资料可以参考 https://developer.qualcomm.com/software/qualcomm-neural-processing-sdk

第 6 章 端侧模型部署框架

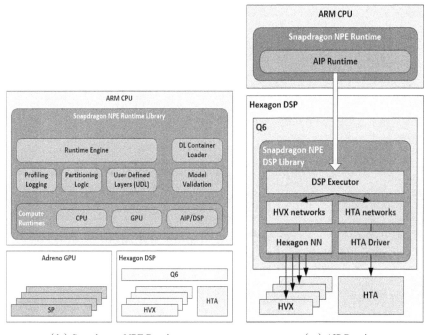

（a）SNPE 使用流程

（b）Snapdragon NPE Runtime　　　（c）AIP Runtime

图 6-8　SNPE 执行流程[1]

QNN（Qualcomm® AI Engine Direct SDK）是 Qualcomm 新推出的模型部署方案。它提供了较底层的统一 API（lower-level，unified APIs），更接近处理器底层去提高 AI 模型的性能，其架构如图 6-9 所示。

---

[1]　图来源：https://developer.qualcomm.com

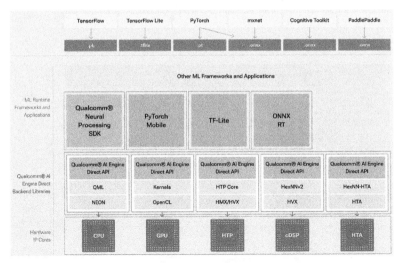

图 6-9 QNN 架构

（其中 HTP 代表 Hexagon processor with fused AI accelerator architecture；cDSP 代表 Hexagon processor without fused AI accelerator architecture；HTA 代表 Legacy, standalone tensor accelerator）[1]

6.10 MNN

MNN 是阿里开源的深度学习移动端推理框架，其支持的算子（Supported Ops）包括 CPU、Metal、Vulkan、OpenCL 等。其架构如图 6-8 所示，支持 TensorFlow、Caffe、ONNX 等模型格式转换。

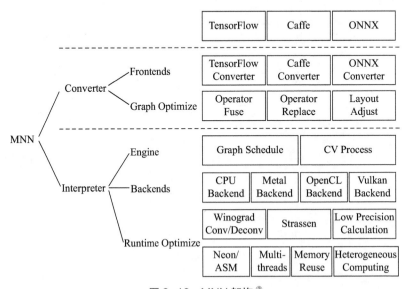

图 6-10 MNN 架构[2]

---

① 图来源：https://developer.qualcomm.com/software/qualcomm-ai-engine-direct-sdk
② 图来源：https://github.com/alibaba/MNN.git

6.11 MediaPipe

MediaPipe 是 Google 开源的跨平台（Cross-platform）和可定制的服务于直播和流媒体的机器学习应用框架。通过 MediaPipe 可以轻松实现自拍分割（Selfie Segmentation）、人脸检测（Face Detection）、人脸网格（Face Mesh）、手势跟踪（Hand Tracking）、人体姿态检测与跟踪（Human Pose Detection and Tracking）、3D 目标检测（3D Object Detection）等，如图 6-11 所示。

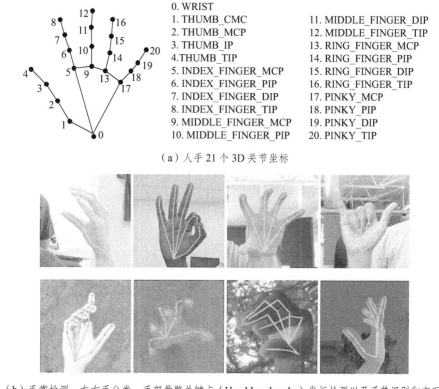

（a）人手 21 个 3D 关节坐标

（b）手掌检测、左右手分类、手部骨骼关键点（Hand Landmarks）坐标检测以及手势识别和交互

图 6-11 MediaPipe[①]

6.12 NeuroPilot

NeuroPilot 是 MediaTek 推出的实现在边缘端侧设备上开发和部署 AI 应用程序的解决方案。[②]

---

① 图来源：https://google.github.io/mediapipe/solutions/hands
② 参考资料：https://developer.mediatek.com/ai/6423c8b6c612745b3a4baa2c.html

第 7 章

ARM 处理器 OpenCV 编程

OpenCV（Open Source Computer Vision Library，开源计算机视觉库）是 Intel 完全开源的计算机视觉库，其生命力非常旺盛，版本一直迭代——"在和我们一起成长"。OpenCV 有不同的功能接口，通过学习它的各个模块的使用和研究底层代码实现，可以快速支持我们的产品开发，进而一步步去积累经验和增长知识。OpenCV 的 Python 版本和 C/C++ 版本都是很好的入门版本。OpenCV 中的功能模块非常多，涵盖 CV 的各个场景领域（OpenCV 3.3 版本开始支持 DNN 模块），表 7-1 所示为 OpenCV 的模块结构：

表 7-1 OpenCV 模块结构[①]

	子模块	说明
基本模块	core	核心库模块，基本数据结构、操作函数
	imgcodecs	图像读写模块，图像文件读取、保存
	imgproc	图像处理模块，图像滤波、几何变换、直方图、特征检测与目标检测
其他模块	objdetect	目标检测模块
	dnn	深度神经网络模块
	highgui	可视化模块，High-level GUI
	calib3d	相机标定与三维重建模块
	……	……

OpenCV 源码的最底层模块是 HAL（Hardware Acceleration Layer，硬件加速层），它是基于底层硬件开发并且优化的 OpenCV 算子。HAL 涉及的技术有 NEON、SSE（Streaming SIMD Extensions，单指令多数据流扩展）、IPP（Intel Integrated Performance Primitives，Intel IPP）、OpenCL、CUDA 等。在实际项目开发中往往只依赖其中部分模块和代码，所以先做模块配置，再编译 OpenCV 库。

7.1 ARM 平台移植 OpenCV 库

OpenCV 在 ARM 平台的移植，所涉及的内容包括库交叉编译和 ARM 平台测试。在这里我们以 RV1126 平台为例讨论如何移植。

运行以下命令：

git clone -b 4.9.0 https://github.com/opencv/opencv.git

下载指定分支的 OpenCV 工程源码。

[①] 更多资料可参考 https://docs.opencv.org/4.9.0/index.html

接下来使用 cmake-gui 工具来构建编译的文件，如图 7-1 所示配置代码路径和编译输出路径，然后点击 configure。

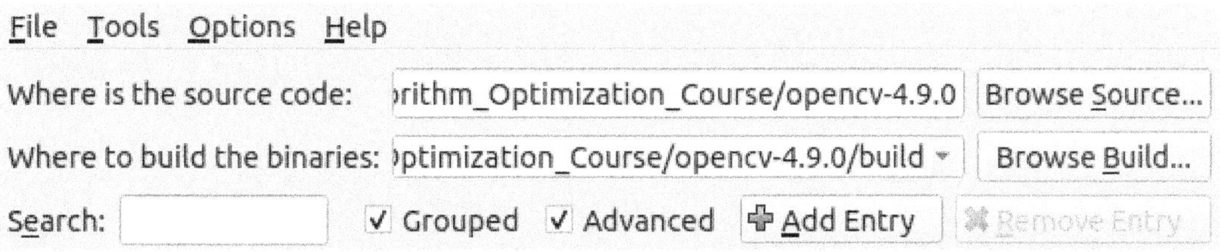

图 7-1　配置代码路径和编译输出路径

继续如图 7-2 所示配置，点击【Next】。

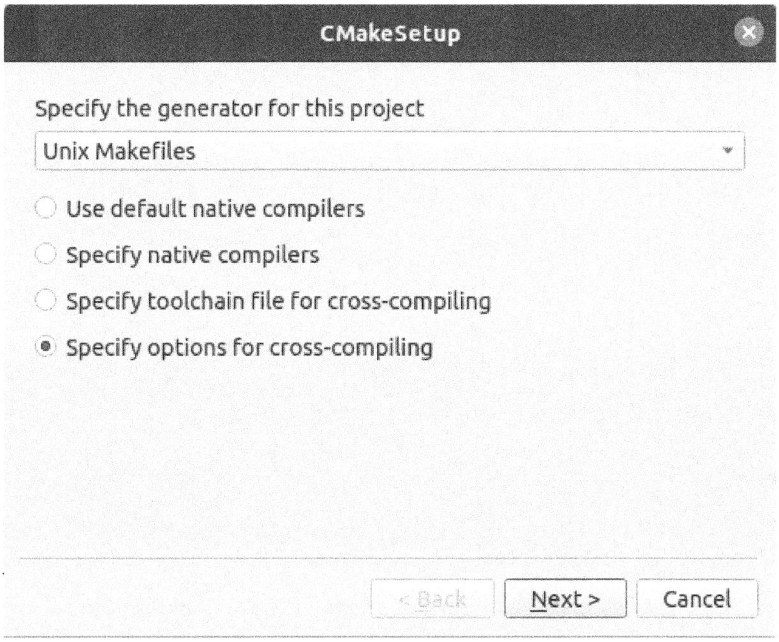

图 7-2　交叉编译工具配置

接下来配置 C 和 C++ 编译器，分别为：/opt/atk-dlrv1126-toolchain/usr/bin/arm-linux-gnueabihf-gcc 和 /opt/atk-dlrv1126-toolchain/usr/bin/arm-linux-gnueabihf-g++，如图 7-3 所示，点击【Finish】。

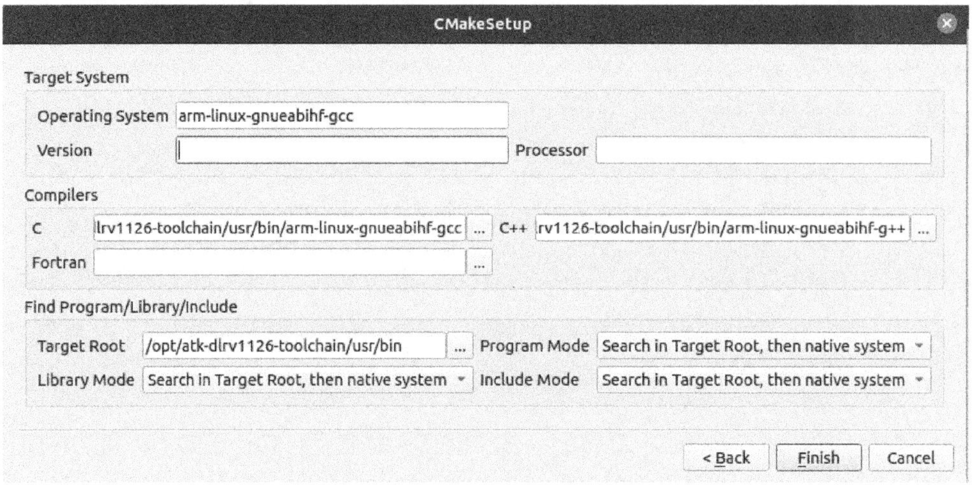

图 7-3 配置 C 和 C++ 编译器

勾选 CMake 界面的【Grouped】和【Advanced】选项，如图 7-4 所示。

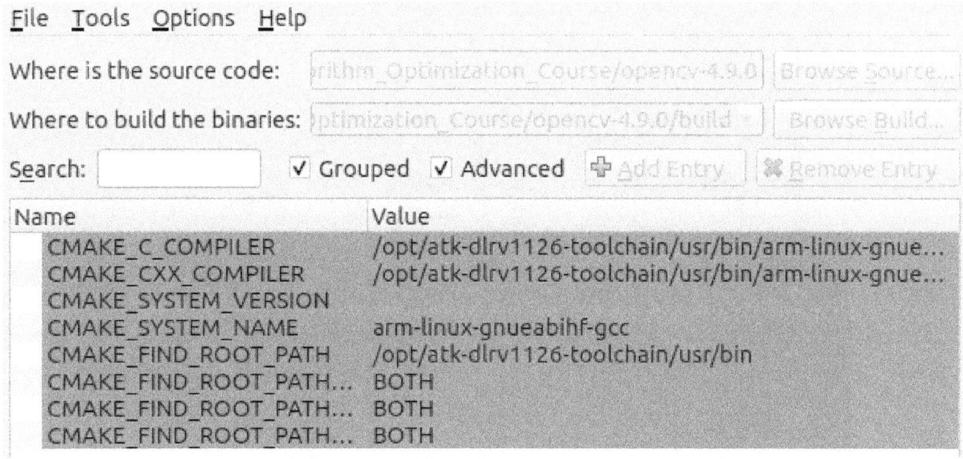

图 7-4 勾选 CMake 界面的【Grouped】和【Advanced】选项

接下来还需要配置的编译选项，分别如下：

（1）配置 CMAKE_INSTALL_PREFIX 指定安装路径，如图 7-5 所示。

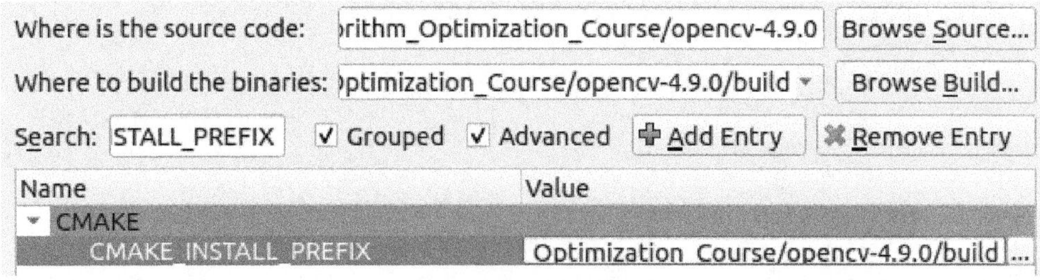

图 7-5 配置安装路径

（2）配置 CMAKE_CXX_FLAGS 链接器选项为"-lpthread -lrt -ldl"，其中"lpthread"代表支持多线程编程（Pthreads），"lrt"代表链接实时库（Realtime），"ldl"代表链接动态加载库（Dynamic Loading），如图 7-6 所示。

图 7-6 配置链接器选项

（3）配置 CMAKE_CXX_FLAGS_DEBUG 为"-lpthread -lrt -ldl"，如图 7-7 所示。

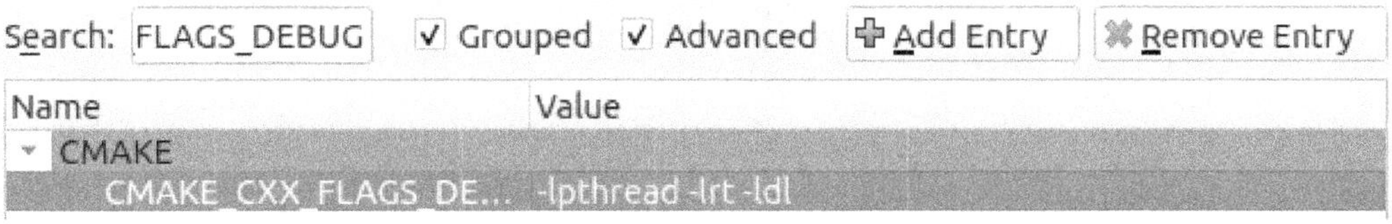

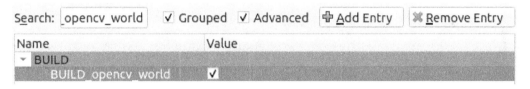

图 7-7 配置 CMAKE_CXX_FLAGS_DEBUG

（4）勾选 BUILD_opencv_world，如图 7-8 所示。

图 7-8 勾选 BUILD_opencv_world

（5）勾选 BUILD_ZLIB，如图 7-9 所示。

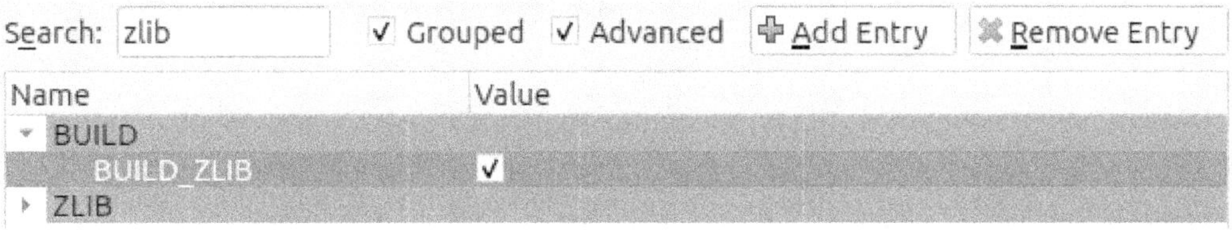

图 7-9 勾选 BUILD_ZLIB

（6）OPENCV_EXTRA_MODULES_PATH 配置，如图 7-10 所示。

下载 opencv_contrib-4.9.0 文件，拷贝到 opencv-4.9.0 文件夹中。

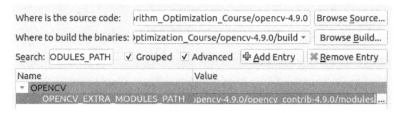

图 7-10　配置 OPENCV_EXTRA_MODULES_PATH

（7）修改 opencv-4.9.0/3rdparty/protobuf/src/google/protobuf/stubs/common.cc 代码，在代码中添加 #define HAVE_PTHREAD，如图 7-11 所示。

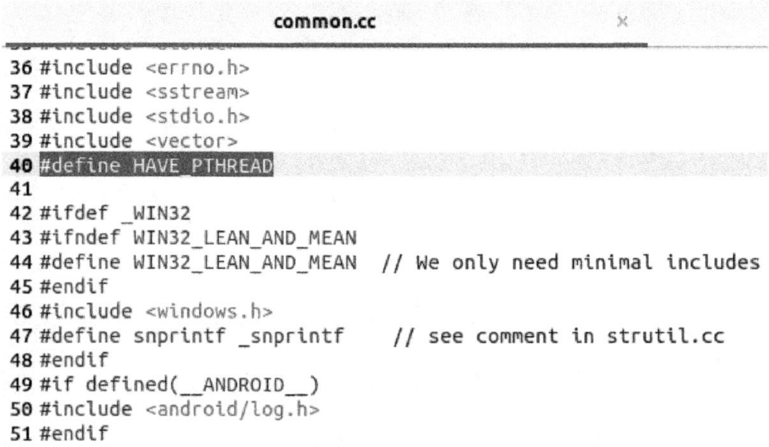

图 7-11　添加 #define HAVE_PTHREAD

点击【configure】，再点击【Generate】。

如果没有出现错误，在终端中运行命令 make -j 8 编译源码。编译成功后，再运行命令 make install，得到库文件和头文件如图 7-12 所示。

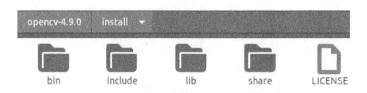

图 7-12　编译安装得到的头文件和库文件

接下来我们使用以下测试代码来验证我们编译的库，完整测试代码如下：

#include <stdlib.h>

#include <iostream>

#include<opencv2/opencv.hpp>

```
using namespace std;
using namespace cv;

int main()
{
  Mat img = imread("test.jpg");
  return 0;
}
```
代码编译和测试方法如下：

$ export PATH=$PATH:/opt/atk-dlrv1126-toolchain/usr/bin

$ export OpenCV_PATH=/******/On-device_AI_Algorithm_Optimization_Course/opencv-4.9.0

$ arm-linux-gnueabihf-g++ threshold.cpp -o test -fPIC -lrt -D_GNU_SOURCE -lpthread -lm -ldl -lopencv_world -I ${OpenCV_PATH}/install/include -L ${OpenCV_PATH}/install/lib

$ opencv-4.9.0/install/lib$ adb push libopencv_world.so /opencv/lib
$ opencv-4.9.0/install/bin$ adb push opencv_* /opencv/bin
在硬件终端配置
export LD_LIBRARY_PATH=/opencv/lib:$LD_LIBRARY_PATH

7.2 OpenCV 库编译错误及解决方法

（1）ADE 文件下载

CMake Error at modules/gapi/cmake/DownloadADE.cmake:23（add_library）:
  No SOURCES given to target: ade
Call Stack（most recent call first）:
  modules/gapi/cmake/init.cmake:20（include）
  cmake/OpenCVModule.cmake:298（include）
  cmake/OpenCVModule.cmake:361（_add_modules_1）
  cmake/OpenCVModule.cmake:408（ocv_glob_modules）
  CMakeLists.txt:1032（ocv_register_modules）

解决方法：

分析 build 中的 CMakeDownloadLog.txt 日志，发现问题：

#cmake_download" /******/opencv-4.9.0/.cache/ade/***-v0.1.2d.zip" " https://github.

com/opencv/ade/archive/v0.1.2d.zip"

#try 1

******

TCP_NODELAY set

connect to *** port *** failed: Connection timed out

我们可以直接下载文件 https://github.com/opencv/ade/archive/v0.1.2d.zip，或者修改配置文件，取消勾选这个模块。

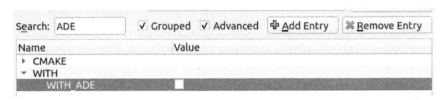

图 7-13　ADE 配置选项

（2）CMAKE_SYSTEM_PROCESSOR

CMake Warning at cmake/OpenCVDetectCXXCompiler.cmake:80（message）：

　OpenCV: CMAKE_SYSTEM_PROCESSOR is not defined. Perhaps CMake toolchain is broken

Call Stack（most recent call first）：

　CMakeLists.txt:174（include）

解决方法：

修改 cmake/OpenCVDetectCXXCompiler.cmake 文件，如下图所示：

```
 87 message(STATUS "Detected processor: ${CMAKE_SYSTEM_PROCESSOR}")
 88 if(OPENCV_SKIP_SYSTEM_PROCESSOR_DETECTION)
 89   # custom setup: required variables are passed through cache / CMake's command-line
 90 elseif(CMAKE_SYSTEM_PROCESSOR MATCHES "amd64.*|x86_64.*|AMD64.*")
 91   set(X86_64 1)
 92 elseif(CMAKE_SYSTEM_PROCESSOR MATCHES "i686.*|i386.*|x86.*")
 93   set(X86 1)
 94 elseif(CMAKE_SYSTEM_PROCESSOR MATCHES "^(aarch64.*|AARCH64.*|arm64.*|ARM64.*)")
 95   set(AARCH64 1)
 96 elseif(CMAKE_SYSTEM_PROCESSOR MATCHES "^(arm.*|ARM.*)")
 97   set(ARM 1)
 98 elseif(CMAKE_SYSTEM_PROCESSOR MATCHES "^(powerpc|ppc)64le")
 99   set(PPC64LE 1)
100 elseif(CMAKE_SYSTEM_PROCESSOR MATCHES "^(powerpc|ppc)64")
101   set(PPC64 1)
102 elseif(CMAKE_SYSTEM_PROCESSOR MATCHES "^(mips.*|MIPS.*)")
103   set(MIPS 1)
104 elseif(CMAKE_SYSTEM_PROCESSOR MATCHES "^(riscv.*|RISCV.*)")
105   set(RISCV 1)
106 elseif(CMAKE_SYSTEM_PROCESSOR MATCHES "^(loongarch.*|LOONGARCH.*)")
107   set(LOONGARCH64 1)
108 else()
109   #if(NOT OPENCV_SUPPRESS_MESSAGE_UNRECOGNIZED_SYSTEM_PROCESSOR)
110   # message(WARNING "OpenCV: unrecognized target processor configuration")
111   #endif()
112   set(CMAKE_SYSTEM_PROCESSOR_arm)
113 endif()
```

图 7-14　修改 cmake/OpenCVDetectCXXCompiler.cmake

（3）operator '&&' has no right operand

Scanning dependencies of target libtiff

［3%］Building C object 3rdparty/libtiff/CMakeFiles/libtiff.dir/tif_aux.c.obj

In file included from /opt/atk-dlrv1126-toolchain/prebuilts/gcc/linux-x86/arm/gcc-arm-8.3-2019.03-x86_64-arm-linux-gnueabihf/arm-linux-gnueabihf/libc/usr/include/fcntl.h:25，

from /******/opencv-4.9.0/3rdparty/libtiff/tiffiop.h:34，

from /******/opencv-4.9.0/3rdparty/libtiff/tif_aux.c:30:

/opt/atk-dlrv1126-toolchain/prebuilts/gcc/linux-x86/arm/gcc-arm-8.3-2019.03-x86_64-arm-linux-gnueabihf/arm-linux-gnueabihf/libc/usr/include/features.h:363:52: error: operator '&&' has no right operand

　　#if defined _FILE_OFFSET_BITS && _FILE_OFFSET_BITS == 64

解决方法：

将 opencv-4.9.0/3rdparty/libtiff/tif_config.h.cmake.in 文件中的配置代码注释掉：

/*#define _FILE_OFFSET_BITS @FILE_OFFSET_BITS@*/

（4）boostdesc_bgm.i

［64%］Building CXX object modules/world/CMakeFiles/opencv_world.dir/******/opencv_contrib-3.4.3/modules/xfeatures2d/src/boostdesc.cpp.obj

/rk1126/opencv_contrib-3.4.3/modules/xfeatures2d/src/boostdesc.cpp:653:20: fatal error: boostdesc_bgm.i: No such file or directory

　　　#include "boostdesc_bgm.i"

compilation terminated.

make［2］: ***［modules/world/CMakeFiles/opencv_world.dir/build.make:7158: modules/world/CMakeFiles/opencv_world.dir/******/rk1126/opencv_contrib-3.4.3/modules/xfeatures2d/src/boostdesc.cpp.obj］Error 1

make［1］: ***［CMakeFiles/Makefile2:3130: modules/world/CMakeFiles/opencv_world.dir/all］Error 2

make: ***［Makefile:163: all］Error 2

解决方法：

我们在 github.com/opencv/opencv_contrib/issues/1301 可以找到相同的问题。

在 /opencv/build 文件夹中可以查询日志文件 CMakeDownloadLog.txt，下载以下文件：

boostdesc_bgm_bi.i	boostdesc_binboost_128.i	vgg_generated_48.i
boostdesc_bgm_hd.i	boostdesc_binboost_256.i	vgg_generated_64.i
boostdesc_bgm.i	boostdesc_lbgm.i	vgg_generated_80.i

boostdesc_binboost_064.i vgg_generated_120.i

拷贝到 opencv_contrib*.*.*/modules/xfeatures2d/src/ 路径下，然后输入命令 make clean 和 make -j 8。

（5）png_init_filter_functions_neon

［71%］Linking CXX executable ../../bin/opencv_test_videostab

/opt/atk-dlrv1126-toolchain/prebuilts/gcc/linux-x86/arm/gcc-arm-8.3-2019.03-x86_64-arm-linux-gnueabihf/bin/../lib/gcc/arm-linux-gnueabihf/8.3.0/../../../../arm-linux-gnueabihf/bin/ld: ../../lib/libopencv_world.so: undefined reference to `png_init_filter_functions_neon'

解决方法：

将文件 /opencv/3rdparty/libpng/pngpriv.h 中代码

if(defined（__ARM_NEON__）|| defined（__ARM_NEON））&& \
　 defined（PNG_ALIGNED_MEMORY_SUPPORTED）
#　　define PNG_ARM_NEON_OPT 2
else
#　　define PNG_ARM_NEON_OPT 0
endif
#endif

修改为：

/*# if(defined（__ARM_NEON__）|| defined（__ARM_NEON））&& */
#　if defined（PNG_ARM_NEON）&&（defined（__ARM_NEON__）|| defined（__ARM_NEON））&& \
　 defined（PNG_ALIGNED_MEMORY_SUPPORTED）
　　 define PNG_ARM_NEON_OPT 2
else
#　　define PNG_ARM_NEON_OPT 0
endif
#endif

（6）cv::dnn::opt_NEON::convBlock_F32

../../lib/libopencv_world.so: undefined reference to `cv::dnn::opt_NEON::convBlock_F32（int, float const*, float const*, float*, int, bool, int, int, int）'

解决方法：

修改文件 opencv-4.9.0/CMakeLists.txt，在其中添加配置：SET（ARM 1）和 SET（ENABLE_NEON 1）。

7.3 ARM 平台 C/C++ 图像处理实例

本节主要讨论在 ARM 平台中实现图像处理，其中的测试用例是基于 RV1126 芯片，代码的编译方法如下：

arm-linux-gnueabihf-g++ test.cpp -o test -fPIC -lrt -D_GNU_SOURCE -lpthread -lm -ldl -lopencv_world -I ${OpenCV_PATH}/install/include -L ${OpenCV_PATH}/install/lib

如果换成其他测试平台，需要适配对应的编译器，代码部分基本不用修改。

在 RV1126 平台需要配置库文件环境变量：export LD_LIBRARY_PATH=/opencv/lib:$LD_LIBRARY_PATH，这样程序运行才能找到 OpenCV 的库，或者直接将库文件拷贝到系统库文件。

我们可以使用 file test 命令查看编译得到的可执行文件是否为 ARM 平台可执行文件，如图 7-15 所示。

```
lgorithm_Optimization_Course/test_opencv$ file test
test: ELF 32-bit LSB executable, ARM, EABI5 version 1 (SYSV), dynamically linked
, interpreter /lib/ld-linux-armhf.so.3, for GNU/Linux 3.2.0, with debug_info, no
t stripped
```

图 7-15 查看编译得到的可执行文件属性

使用命令行 readelf -a test | grep NEEDED 查看可执行程序中调用库的依赖关系，如图 7-16 所示，通过这个工具可以辅助分析运行错误，比如在运行中出现问题：./test: /lib/libstdc++.so.6: no version information available（required by ./test），通过工具可以看到依赖关系如图（b）与图（a）不同。

```
0x00000001 (NEEDED)                     Shared library: [librt.so.1]
0x00000001 (NEEDED)                     Shared library: [libpthread.so.0]
0x00000001 (NEEDED)                     Shared library: [libdl.so.2]
0x00000001 (NEEDED)                     Shared library: [libopencv_world.so]
0x00000001 (NEEDED)                     Shared library: [libstdc++.so.6]
0x00000001 (NEEDED)                     Shared library: [libm.so.6]
0x00000001 (NEEDED)                     Shared library: [libgcc_s.so.1]
0x00000001 (NEEDED)                     Shared library: [libc.so.6]
```

（a）

```
lgorithm_Optimization_Course/Chapter_8_testOpenCV$ readelf -a test | grep NEEDED
0x00000001 (NEEDED)                     Shared library: [libopencv_world.so]
0x00000001 (NEEDED)                     Shared library: [libstdc++.so.6]
0x00000001 (NEEDED)                     Shared library: [libgcc_s.so.1]
0x00000001 (NEEDED)                     Shared library: [libc.so.6]
0x00000001 (NEEDED)                     Shared library: [ld-linux-armhf.so.3]
```

（b）

图 7-16 调用库的依赖关系

7.3.1 图像浅拷贝和深拷贝

OpenCV C/C++ 中图像浅拷贝（shallow copy）属于指针拷贝，其特点是没有申请新的内存空间，对浅拷贝后进行的操作会同样作用于原图像。

而图像深拷贝（deep copy）属于内容拷贝，其特点是申请新的内存空间，对深拷贝后进行的操作和原图像没有关联，其方法包括 copyTo（）和 clone（）。

关于图像浅拷贝和深拷贝完整测试代码如下：

```cpp
#include <stdlib.h>
#include <iostream>
#include<opencv2/opencv.hpp>
#include <sys/time.h>

using namespace std;
using namespace cv;

int main（）
{
    Mat img = imread（"test0.png"）;
    if（img.empty（））{
        printf（"could not load the image \n"）;
        return -1;
    }
    // shallow copy
    Mat img_1（img）;
    Mat img_2 = img;
cv::rectangle（img_1, cv::Point（208, 198）, cv::Point（727,
            528）, cv::Scalar（0, 0, 255, 0), 5）;
    cv::imwrite（"img_1.jpg", img_1）;
cv::rectangle（img_2, cv::Point（208, 198）, cv::Point（727,
            528）, cv::Scalar（0, 0, 255, 0), 5）;
    cv::imwrite（"img_2.jpg", img_2）;

    // deep copy
    Mat img_3 = img.clone（）;
    Mat img_4;
    img.copyTo（img_4）;

cv::rectangle（img_3, cv::Point（208, 198）, cv::Point（727,
            528）, cv::Scalar（0, 0, 255, 0), 5）;
    cv::imwrite（"img_3.jpg", img_3）;
```

```
    cv::rectangle（img_4, cv::Point（208，198），cv::Point（727，
            528），cv::Scalar（0，0，255，0），5）;
    cv::imwrite（"img_4.jpg"，img_4）;
    return 0;
}
```

代码运行结果如图 7-17 所示：

图 7-17　图像浅拷贝和深拷贝测试结果

7.3.2　颜色空间转换

OpenCV C/C++ 可实现图像颜色空间转换，测试代码如下：

```
#include <stdlib.h>
#include <iostream>
#include <opencv2/opencv.hpp>

using namespace std;
using namespace cv;

int main（）
{
    Mat img = imread（"test.jpg"）;
    Mat gray = Mat::zeros（img.size（），CV_8UC1）;
    cvtColor（img, gray, COLOR_BGR2GRAY）;
    imwrite（"gray.jpg"，gray）;
    return 0;
}
```

代码编译和测试方法如下：

arm-linux-gnueabihf-g++ test.cpp -o test -fPIC -lrt -D_GNU_SOURCE -lpthread -lm -ldl -lopencv_world -I ${OpenCV_PATH}/opencv-4.9.0/install/include -L ${OpenCV_PATH}/opencv-4.9.0/install/lib

opencv-4.9.0/install/lib$ adb push libopencv_world.so /opencv/lib

opencv-4.9.0/install/bin$ adb push opencv_* /opencv/bin

export LD_LIBRARY_PATH=/opencv/lib:$LD_LIBRARY_PATH

代码在RV1126平台中运行结果如图7-18所示：

图7-18　颜色空间转换

7.3.3　图像二值化

图像二值化（Image Binarization）是通过单通道图像上的每一个点像素值与特定阈值对比后转换为0或255（或者0或1），进而实现凸显图像中目标轮廓的功能。如图7-19所示，通过图像二值化处理后可以明显看到硬币凸显的轮廓细节。

 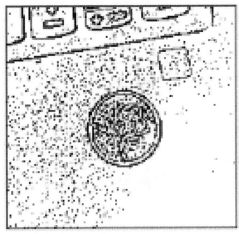

图7-19　图像二值化效果（左图为原图，右图为二值化处理效果）

图像二值化算法如表 7-2 所示:

表 7-2 图像二值化算法[①]

图像二值化算法	代码摘录
固定阈值二值化(Simple Thresholding)	cv.THRESH_BINARY cv.THRESH_BINARY_INV cv.THRESH_TRUNC cv.THRESH_TOZERO cv.THRESH_TOZERO_INV
	import cv2 as cv import numpy as np img = cv.imread('test.png', 0) ret, thresh1 = cv.threshold(img, 127, 255, cv.THRESH_BINARY) ret, thresh2 = cv.threshold(img, 127, 255, cv.THRESH_BINARY_INV) ret, thresh3 = cv.threshold(img, 127, 255, cv.THRESH_TRUNC) ret, thresh4 = cv.threshold(img, 127, 255, cv.THRESH_TOZERO) ret, thresh5 = cv.threshold(img, 127, 255, cv.THRESH_TOZERO_INV)
自适应阈值二值化(Adaptive Thresholding)	cv.ADAPTIVE_THRESH_MEAN_C cv.ADAPTIVE_THRESH_GAUSSIAN_C
	import cv2 as cv import numpy as np img = cv.imread('test.png', 0) img = cv.medianBlur(img, 5) ret, th1 = cv.threshold(img, 127, 255, cv.THRESH_BINARY) th2 = cv.adaptiveThreshold(img, 255, cv.ADAPTIVE_THRESH_MEAN_C, cv.THRESH_BINARY, 11, 2) th3 = cv.adaptiveThreshold(img, 255, cv.ADAPTIVE_THRESH_GAUSSIAN_C, cv.THRESH_BINARY, 11, 2)

---

① 参考 https://docs.opencv.org/5.x/d7/d4d/tutorial_py_thresholding.html

续表

图像二值化算法	代码摘录

基于直方图二值化（Otsu's Binarization）	cv.THRESH_OTSU
	import cv2 as cv
	import numpy as np
	img = cv.imread（'test.png'，0）
	# global thresholding
	ret1，th1 = cv.threshold（img，127，255，cv.THRESH_BINARY）
	# Otsu's thresholding
	ret2，th2 = cv.threshold（img，0，255，cv.THRESH_BINARY + cv.THRESH_OTSU）
	# Otsu's thresholding after Gaussian filtering
	blur = cv.GaussianBlur（img，(5，5)，0）
	ret3，th3 = cv.threshold（blur，0，255，cv.THRESH_BINARY + cv.THRESH_OTSU）

在ARM平台中，通过OpenCV C/C++实现图像二值化处理的完整测试代码如下：

```cpp
#include <stdlib.h>
#include <iostream>
#include<opencv2/opencv.hpp>

using namespace std;
using namespace cv;

int main（）
{
  Mat img = imread（"test1.jpg"）;
  Mat gray_0 = Mat::zeros（img.size（），CV_8UC1）;
  cvtColor（img，gray_0，COLOR_BGR2GRAY）;
  Mat threshold_0; // 二值图
  threshold（gray_0，threshold_0，130，255，THRESH_BINARY）;

  imwrite（"gray_0.jpg"，gray_0）;
  imwrite（"threshold_0.jpg"，threshold_0）;

  cv::Mat gray_1 = imread（"test1.jpg"，CV_8UC1）;
  for（int h = 0; h < gray_1.rows; h++）
  {
```

```
    for( int w = 0; w < gray_1.cols; w++ )
    {
      if( gray_1.at<uint8_t> ( h, w )> 130 ){
        gray_1.at<uint8_t> ( h, w )= 255;
      } else {
        gray_1.at<uint8_t> ( h, w )= 0;
      }
    }
  }
  imwrite ( ″threshold_1.jpg″, gray_1 );

  return 0;
}
```

代码运行结果如图 7-20 所示：

图 7-20　图像二值化结果（照片拍摄于美丽的大兴安岭，赏梅花鹿的灵动之美，"呦呦鹿鸣，食野之苹"。）

7.3.4　图像翻转 / 旋转 / 缩放 / 裁剪

OpenCV C/C++ 可轻松实现图像的翻转、旋转、缩放和裁剪，其完整测试代码如下：

```
#include <stdlib.h>
#include <iostream>
#include<opencv2/opencv.hpp>

using namespace std;
using namespace cv;

int main ( )
{
```

```cpp
Mat img = imread("test.jpg"); // 图像文件读取
Mat gray = Mat::zeros(img.size(), CV_8UC1);
cvtColor(img, gray, COLOR_BGR2GRAY);
imwrite("gray.jpg", gray); // 图像文件保存

// 翻转
cv::Mat flip_img;
cv::flip(img, flip_img, -1); // 1: 水平翻转; 0: 垂直翻转; -1: 水平垂直翻转
imwrite("flip_img.jpg", flip_img);

// 旋转
cv::Mat rotate_img;
cv::rotate(img,
rotate_img,
cv::ROTATE_90_COUNTERCLOCKWISE);
// cv2.ROTATE_90_CLOCKWISE: 顺时针旋转 90°; cv2.ROTATE_180:
// 旋转 180°; cv2.ROTATE_90_COUNTERCLOCKWISE: 逆时针旋转 90°
imwrite("rotate_img.jpg", rotate_img);

// 缩放
cv::Mat resize_img;
cv::resize(img, resize_img, cv::Size(320, 320));
// (width, height)
imwrite("resize_img.jpg", resize_img);
/*
int h = img.rows;
int w = img.cols;
Size dsize = Size(round(0.5 * w), round(0.5 * h));
resize(src, resize_img, dsize, 0, 0, INTER_AREA);
*/

// 裁剪
cv::Rect rect(150, 300, 90, 150); // x, y, width, height
cv::Mat roi_img = img(rect);
cv::imwrite("roi_img.jpg", roi_img);
```

```
    return 0;
}
```

代码运行结果如图 7-21 所示：

翻转　　　　　　　　　　　旋转

缩放　　　　　　　　　　　裁剪

图 7-21　图像翻转 / 旋转 / 缩放 / 裁剪结果

7.3.5　二维码检测和解码

OpenCV C/C++ 可实现二维码的检测与解码。在 OpenCV 头文件中使用命令 grep -r -w detectAndDecode . –n，查找如下：

opencv-4.9.0/install$ grep -r -w detectAndDecode . -n

./include/opencv2/objdetect/graphical_code_detector.hpp:45:　　CV_WRAP std::string detectAndDecode（InputArray img, OutputArray points = noArray（），

./include/opencv2/wechat_qrcode.hpp:44:　　* To simplify the usage, there is a only API: detectAndDecode

./include/opencv2/wechat_qrcode.hpp:51:　　CV_WRAP std::vector<std::string> detectAndDecode（InputArray img, OutputArrayOfArrays points = noArray（））；

可以发现 OpenCV 中支持二维码检测和解码功能的头文件有 graphical_code_detector.

hpp 和 wechat_qrcode.hpp。第一个头文件，其中定义了一个类 GraphicalCodeDetector，detectAndDecode 是类的一个方法，如图 7-22 所示。

```
39    /** @brief Both detects and decodes graphical code
40
41    @param img grayscale or color (BGR) image containing graphical code.
42    @param points optional output array of vertices of the found graphical code quadrangle,
   will be empty if not found.
43    @param straight_code The optional output image containing binarized code
44    */
45    CV_WRAP std::string detectAndDecode(InputArray img, OutputArray points = noArray(),
46                                       OutputArray straight_code = noArray()) const;
```

图 7-22　类 GraphicalCodeDetector 和类的方法 detectAndDecode [1]

文件 ./include/opencv2/objdetect.hpp:770:class CV_EXPORTS_W_SIMPLE QRCodeDetector: public GraphicalCodeDetector 中的类 QRCodeDetector 继承 GraphicalCodeDetector，于是类 QRCodeDetector 继承了方法 detectAndDecode，实现二维码检测和解码的代码如下：

QRCodeDetector QRdetecter;

string info = QRdetecter.detectAndDecode（img, points, straight_qrcode）;

如图 7-23 所示，实现代码在测试图片中检测到的二维码的四个顶点坐标为（178，27）、（358，163）、（229，350）、（41，209）。straight_qrcode 为输出的先矫正后二值化处理的二维码图像。测试图片的尺寸 width 为 429 个像素，height 为 377 个像素，在 RV1126 平台时耗 263 ms。

图 7-23　二维码检测、解码和矫正结果（右图为矫正结果）

完整的测试代码如下：

#include <stdlib.h>

#include <iostream

---

[1] 内容来源：./include/opencv2/objdetect/graphical_code_detector.hpp:45

```cpp
#include<opencv2/opencv.hpp>
#include <sys/time.h>

using namespace std;
using namespace cv;

int main()
{
    Mat img = imread("qrcode.png");
    Mat straight_qrcode;
    vector<Point> points;
    //std::vector<cv::Point> points;

    QRCodeDetector QRdetecter;
    struct timeval time_begin, time_end;
    //#include <sys/time.h>
    gettimeofday(&time_begin, NULL);
    string info = QRdetecter.detectAndDecode(img, points, straight_qrcode);
    gettimeofday(&time_end, NULL);
    int time = (1000000 * (time_end.tv_sec - time_begin.tv_sec)
        + (time_end.tv_usec - time_begin.tv_usec)) / 1000;
    printf("cost: %d ms\n", time)

    Point point1 = Point(0, 30);
    putText(img, info, point1, FONT_HERSHEY_SIMPLEX, 0.6, Scalar(0,
            0, 255), 1, 8);
for(int j = 0; j < points.size(); j++)
{
        printf(" points %d, %d\n", points[j].x, points[j].y);
        if(j == 3)
            line(img, points[j], points[0],
Scalar(0, 0, 255), 3);
        else
            line(img, points[j], points[j + 1],
Scalar(0, 0, 255), 3);
```

```
    }

    cv::imwrite("img.jpg", img);
    cv::imwrite("straight_qrcode.jpg", straight_qrcode);
    return 0;
}
```

第 8 章

Neon 指令集加速算法和算子底层指令加速

Neon 指令是 ARM 处理器的 SIMD 指令集，类似 X86 的 SSE2，它适用于 ARM 中 Cortex-A 和 Cortex-R 系列处理器。Neon Intrinsics 是官方提供的 Neon C 和 C++ 内联函数，可实现底层 Neon 汇编的调用，相比较直接使用汇编更加容易编写和维护。Neon 内联函数接口和数据类型定义在 arm_neon.h 头文件中。本章将讨论如何使用 Neon 内联函数进行 Neon 编程优化算法。

8.1 Neon 寄存器和数据类型

Neon 寄存器，也叫向量寄存器或向量，它包括 16 个 128 bit 四字寄存器 Q0—Q15 或 32 个 64 bit 双字寄存器 D0—D31，其中一个字为 32 bit。

Q0—Q15 中每一个寄存器映射到一对 D 寄存器，比如 Q0 映射 D0 和 D1 寄存器，Q1 映射 D2 和 D3 寄存器，Q2 映射 D4 和 D5 寄存器，以此类推，如图 8-1 所示。

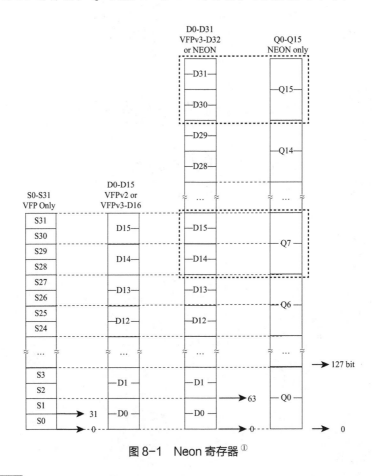

图 8-1 Neon 寄存器[①]

---

① 资料来源：https://developer.arm.com/documentation/102474/latest https://developer.arm.com/documentation/den0018/latest

图来源：NEON Programmer's Guide Version: 1.0

ARM 处理器的基本数据类型有整数、浮点数、poly，具体如表 8-1 所示。

表 8-1　ARM 基本数据类型

整数	int8、int16、int32、int64
	uint8、uint16、uint32、uint64
浮点数	float16、float32
多项式	poly8、poly16

ARM 处理器中的 Neon 数据类型如表 8-2 所示。

表 8-2　Neon 数据类型

	8 bit	16 bit	32 bit	64 bit
无符号整型 （Unsigned Integer）	U8	U16	U32	U64
	$[0, 2^8)$	$[0, 2^{16})$	$[0, 2^{32})$	$[0, 2^{64})$
有符号整型 （Signed Integer）	S8	S16	S32	S64
	$[-2^7, 2^7)$	$[-2^{15}, 2^{15})$	$[-2^{31}, 2^{31})$	$[-2^{63}, 2^{63})$
未指定类型的整数 （Integer of Unspecified Type）	I8	I16	I32	I64
浮点型 （Floating-point Number）	not available	F16	F32 or F	not available
{0, 1} 上的多项式 （Polynomial over {0, 1}）	P8	P16	not available	not available

Neon 数据类型和基本数据类型的关系是前者是向量，后者是向量元素的基本类型。如图 8-2 所示为 Neon 内联函数中的数据类型的命名规则。

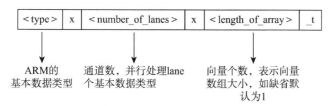

图 8-2　向量数据类型命名规则

当 length_of_array 缺省时，默认为 1，各种数据类型命名示例如图 8-3 所示（arm_neon.h 文件）：

```
typedef __simd64_int8_t int8x8_t;
typedef __simd64_int16_t int16x4_t;
typedef __simd64_int32_t int32x2_t;
typedef __builtin_neon_di int64x1_t;
#if defined (__ARM_FP16_FORMAT_IEEE) || defined (__ARM_FP16_FORMAT_ALTERNATIVE)
typedef __fp16 float16_t;
typedef __simd64_float16_t float16x4_t;
#endif
typedef __simd64_float32_t float32x2_t;
typedef __simd64_poly8_t poly8x8_t;
typedef __simd64_poly16_t poly16x4_t;
#pragma GCC push_options
#pragma GCC target ("fpu=crypto-neon-fp-armv8")
typedef __builtin_neon_poly64 poly64x1_t;
#pragma GCC pop_options
typedef __simd64_uint8_t uint8x8_t;
typedef __simd64_uint16_t uint16x4_t;
typedef __simd64_uint32_t uint32x2_t;
typedef __builtin_neon_udi uint64x1_t;

typedef __simd128_int8_t int8x16_t;
typedef __simd128_int16_t int16x8_t;
typedef __simd128_int32_t int32x4_t;
typedef __simd128_int64_t int64x2_t;
#if defined (__ARM_FP16_FORMAT_IEEE) || defined (__ARM_FP16_FORMAT_ALTERNATIVE)
typedef __simd128_float16_t float16x8_t;
#endif
typedef __simd128_float32_t float32x4_t;
typedef __simd128_poly8_t poly8x16_t;
typedef __simd128_poly16_t poly16x8_t;
#pragma GCC push_options
#pragma GCC target ("fpu=crypto-neon-fp-armv8")
typedef __builtin_neon_poly64 poly64x2_t __attribute__ ((__vector_size__ (16)));
#pragma GCC pop_options

typedef __simd128_uint8_t uint8x16_t;
typedef __simd128_uint16_t uint16x8_t;
typedef __simd128_uint32_t uint32x4_t;
typedef __simd128_uint64_t uint64x2_t;

typedef float float32_t;
```

图 8-3　向量数据类型命名示例（length_of_array 缺省）

当 length_of_array 不缺省，被定义的数据类型也叫向量数组（数组的元素是向量），它实际上是一个结构体，结构体中定义向量，而且每一个向量（vector）对应 1 个 Neon 寄存器。向量中可并行处理的基本数据类型的元素个数叫通道数（lanes），如下文定义的向量数组。

typedef struct int8x16x2_t

{

　int8x16_t val［2］;

} int8x16x2_t;

其中 int8x16x2_t.val［0］和 int8x16x2_t.val［1］为 2 个向量（对应 2 个 Neon 寄存器），每个向量包含 16 个元素，且元素类型为 int8。

8.2 Neon 指令类型

Neon 指令按照操作数类型分为以下几种：

（1）正常指令（Normal Instructions）

对任何向量类型进行操作，并生成与操作数向量相同大小和通常相同类型的结果向量。例如 Neon 汇编指令：VADD.I16 Q0，Q1，Q2。

（2）长指令（Long Instructions）

如图 8-4 所示，对双字向量操作数执行运算，并生成四字向量结果。所生成的元素通常是操作数元素宽度的两倍，并属于同一类型。例如指令：VADDL.S16 Q0，D2，D3。

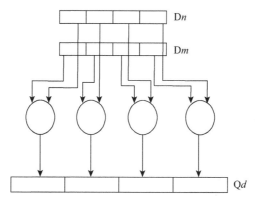

图 8-4　Neon 长指令（操作数 Dn 和 Dm 均是双字向量，生成的结果 Qd 是四字向量）[①]

（3）宽指令（Wide Instructions）

如图 8-5 所示，一个双字向量操作数和一个四字向量操作数执行运算，生成四字向量结果。例如指令：VADDW.I16 Q0，Q1，D4。

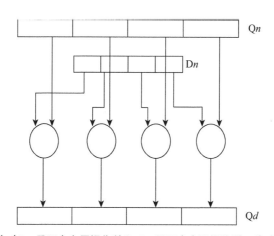

图 8-5　Neon 宽指令（Dn 是双字向量操作数和 Qn 是四字向量操作数，生成结果 Qd 是四字向量）

---

① 图 8-4～图 8-6 来源：NEON Programmer's Guide Version: 1.0

（4）窄指令（Narrow Instructions）

如图 8-6 所示，对四字向量操作数执行运算，并生成双字向量结果。所生成的元素通常是操作数元素宽度的一半。例如指令：VADDHN.I16 D0，Q1，Q2。

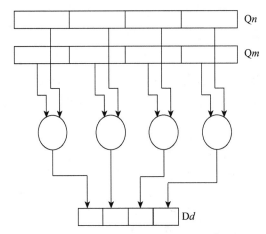

图 8-6　Neon 窄指令（Q*n* 和 Q*m* 是四字向量操作数，生成结果 D*d* 是双字向量）

（5）饱和指令（Saturating Instructions）

在执行运算中，当向量中某个元素的大小超过该数据类型的可表示范围，则会自动限定在该数据类型的表示范围内，防止溢出。例如指令：VQADD、VQSUB、VQABS、VQSHL、VQRSHL。

8.3　Neon 编程方式和内联函数

Neon 编程方式包括 Neon 汇编指令（Neon Assembler）、编译器自动向量化（Auto-vectorization）、Neon 第三方库（Neon-enabled Libraries）以及 Neon 内联函数（Neon Intrinsics）。

8.3.1　Neon 汇编指令

使用 Neon 汇编指令编程，可直接自主分配寄存器，进而能够发挥硬件平台的最大性能，但是代码开发难度大，工程代码可读性低，平台移植性低。鉴于本书篇幅的限制，本书中关于 Neon 汇编指令不做进一步的讨论。

8.3.2　编译器自动向量化

支持自动向量化的编译器工具有 Arm Compiler 6、Arm C/C++ Compiler、LLVM-clang 和 GCC。GCC 工具实现自动向量化的配置参数包括以下所列：

-ftree-vectorize（需要配置 -O1 及以上才能够生效）；
-mfpu = neon（当配置 -O3 参数时 -ftree-vectorize 默认启动）；
-mcpu（-mcpu 参数选项配置如表 8-3 所示）。

表 8-3　-mcpu 参数选项

CPU	option
Cortex-A5	-mcpu = cortex-a5
Cortex-A7	-mcpu = cortex-a7
Cortex-A8	-mcpu = cortex-a8
Cortex-A9	-mcpu = cortex-a9
Cortex-A15	-mcpu = cortex-a15

根据以上参数选项，在 RV1126 平台中使用以下编译命令实现编译自动向量化：
arm-linux-gnueabihf-g++ vadd.cpp -mfpu = neon -mcpu = cortex-a7 -ftree-vectorize -O2 -o test_neon

8.3.3　Neon 第三方库

支持 Neon 的第三方库有 CMSIS-NN、ARM Compute Libray、Ne10、Skia、Libyuv 等。

（1）CMSIS-NN

CMSIS（Cortex Microcontroller Software Interface Standard，Cortex 微控制器软件接口标准）是 ARM 专门针对 Cortex-M 系列提出的一套标准，它是与供应商无关的硬件抽象层。CMSIS 工程代码地址为 https://github.com/ARM-software/CMSIS_5.git，表 8-4 所示为 CMSIS 的代码结构，其中 NN 是 CMSIS 中一个组件，用于支持 ARM Cortex-M 系列处理器内核进行神经网络计算。NN 分为 NNFunctions 和 NNSupportFunctions 两个部分，其中 NNFunctions 包含卷积函数、深度可分离卷积函数、全连接函数、池化函数和激活函数等神经网络中常用的函数，NNSupportFunctions 函数实现关键函数优化，它包括 NNFunctions 中使用的数据转换（用定点 8 或 16 bits 替代浮点计算）和激活功能表（通过查表实现激活函数计算）。

表 8-4　CMSIS 结构[1]

CMSIS	目标处理器	描述
Core（M）	All Cortex-M, SecurCore	Cortex-M 处理器核心和外围设备的标准化 API。包括 Cortex-M4/M7/M33/M35P SIMD 指令的固有功能
Core（A）	Cortex-A5/A7/A9	Cortex-A5/A7/A9 处理器核心和外围设备的标准化 API 和基本运行时系统

---

[1]　摘自：https://arm-software.github.io/CMSIS_5/General/html/index.html

续表

CMSIS	目标处理器	描述
Driver	All Cortex	中间件的通用外围驱动程序接口。将微控制器外围设备与实现例如通信堆栈、文件系统或图形用户界面的中间件连接
DSP	All Cortex-M	DSP 库集合，具有 60 多种功能，适用于各种数据类型：定点（分数 q7、q15、q31）和单精度浮点（32 位）。针对 SIMD 指令集优化的实现可用于 Cortex-M4/M7/M33/M35P
NN	All Cortex-M	收集高效的神经网络内核，以最大限度地提高 Cortex-M 处理器内核的性能和减少内存占用
RTOS v1	Cortex-M0/M0+/M3/M4/M7	实时操作系统的通用 API 以及基于 RTX 的参考实现。它支持可以跨多个 RTOS 系统工作的软件组件
RTOS v2	All Cortex-M，Cortex-A5/A7/A9	使用 Armv8-M 支持、动态对象创建、多核系统规定、二进制兼容接口扩展 CMSIS-RTOS v1
Pack	All Cortex-M，SecurCore，Cortex-A5/A7/A9	描述软件组件、设备参数和评估板支持的传递机制。它简化了软件重用和产品生命周期管理（PLM）
Build	All Cortex-M，SecurCore，Cortex-A5/A7/A9	一组提高生产率的工具、软件框架和工作流程，例如通过持续集成（CI）
SVD	All Cortex-M，SecurCore	可用于在调试器或 CMSIS 核心头文件中创建外设感知的设备的外设描述
DAP	All Cortex	与 CoreSight 调试访问端口接口的调试单元的固件
Zone	All Cortex-M	定义描述系统资源并将这些资源划分为多个项目和执行区域的方法

NN 的框架如图 8-7 所示，从图中可以很好地理解 NNFunctions 和 NNSupportFunctions 的关系。

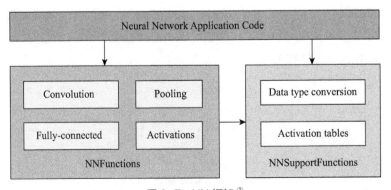

图 8-7　NN 框架[1]

---

[1] 图来源：https://www.keil.com/pack/doc/CMSIS/NN/html/index.html

（2）ARM Compute Libray

ARM Compute Libray（ACL）库的结构如图 8-8 所示，Neon 部分实现 Cortex-A CPU 架构的性能优化，OpenCL 部分实现 Mali GPU 架构的性能优化[①]。

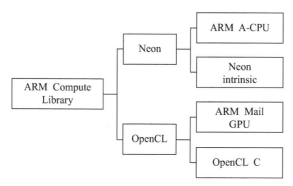

图 8-8　ARM Compute Library

Neon 优化的高性能接口实现 GEMM、Winograd、FFT、滑窗 CNN 和单精度矩阵乘法 SGEMM[②] 等。

ARM Compute Libray 库的编译方法如下：

ComputeLibrary$ scons Werror=0 debug=0 asserts=0 neon=1 opencl=1 os=linux arch=armv7a

……

arm-linux-gnueabihf-ranlib build/tests/framework/libarm_compute_test_framework.a

scons: done building targets.

（3）Ne10

Ne10 是 ARM 官方发布的支持浮点运算、矢量计算、矩阵操作的一个开源库，代码仓库地址为 https://github.com/projectNe10/Ne10。其主要模块包括信号处理 DSP（FFT 快速傅立叶变换、FIR/IIR 滤波等）、数学 math（矢量、矩阵运算等）、图像处理 imagproc（图像缩放、旋转处理）等。以下为 Ne10 的具体模块列表[③]：

Math Functions

　Vector Addition

　Matrix Addition

　Vector Subtraction（w［i］= u［i］- v）

---

① 代码仓库地址：https://github.com/ARM-software/ComputeLibrary

② 参考资料：https://github.com/ARM-software/ComputeLibrary/blob/master/arm_compute/runtime/NEON/NEFunctions.h

③ 来源：http://projectne10.github.io/Ne10/doc/modules.html

Vector Subtraction From (w [i] = v − u [i])

Matrix Subtraction

Vector Multiplication

Vector Multiply-Accumulate

Matrix Multiplication

Matrix-Vector Multiplication

Vector Division

Vector Assignment

Vector Modulus (Length)

Vector Normalization

Vector Element-wise Absolute Value

Vector Dot Product

Vector Cross Product

Matrix Determinant Calculation

Matrix Inversion

Matrix Transposition

Matrix Identity Assignment

Signal Processing Functions

  Complex-to-Complex FFT (Floating & Fixed Point)

  Real-to-Complex FFT (Floating & Fixed Point)

  Finite Impulse Response (FIR) Filters

  Finite Impulse Response (FIR) Decimation

  Finite Impulse Response (FIR) Interpolation

  Finite Impulse Response (FIR) Lattice Filters

  Finite Impulse Response (FIR) Sparse Filters

  Infinite Impulse Response (IIR) Lattice Filters

Image Processing Functions

Image Box Filter (Blur)

Image Resize

Image Rotation

Physics Functions

Collision Detection

(4) Skia

Skia 是开源的 2D 图形库，可用于 web 浏览器和操作系统的图形引擎。

(5) Libyuv

Libyuv 是实现 YUV 缩放和转换等功能的开源库。它的代码仓库地址为：https://developer.arm.com/Architectures/Neon。

8.3.4　Neon 内联函数

Neon 内联函数的命名规则如下：

v <mod> <opname> <shape> <flags>_<type>，其中 opname 代表指令名，type 代表基本数据类型简写。

(1) mod 前缀修饰符表示计算方式，选项有 q、h、d、p 和 r

mod=q 表示饱和计算，如内联函数 int8x8_t vqadd_s8（int8x8_t N, int8x8_t M），用于实现 N 和 M 两个向量的对应元素相加，如果结果溢出，就做饱和处理，其完整测试代码如下：

```cpp
#include <stdio.h>
#include <stdlib.h>
#include <arm_neon.h>
int main()
{
    int8_t vector_0[8]={127, 127, 127, -128, 10, 10, -10, -128};
    int8_t vector_1[8]={127, 127, 10, -128, 20, 100, -100, -2};
    int8_t result[8];

    int8x8_t vec0 = vld1_s8(vector_0);
    int8x8_t vec1 = vld1_s8(vector_1);
    int8x8_t vec_r = vqadd_s8(vec0, vec1);
    vst1_s8(result, vec_r);
    //打印结果
    for(int i = 0; i < 8; i++){
        printf("%d ", result[i]);
    }
    printf("\n");
}
```

编译方法：arm-linux-gnueabihf-g++ vqadd_s8-neon.cpp -o test_neon

完整的测试结果见表 8-5。

表 8-5 mod=q 测试结果

操作数	执行结果
vec0	127, 127, 127, -128, 10, 10, -10, -128
vec1	127, 127, 10, -128, 20, 100, -100, -2
vadd_s8	-2 -2 -119 0 30 110 -110 126
	计算机系统用补码；正数的补码等于原码；负数使用补码表示，即绝对值按位取反后再加 1 127+127 　（01111111）$_b$ 　+（01111111）$_b$ 　=（11111110）$_b$ 　=（-2）$_d$ 127+10 　（01111111）$_b$ 　+（00001010）$_b$ 　=（10001001）$_b$ 　=（-119）$_d$ -128-128 　（10000000）$_b$ 　+（10000000）$_b$ 　=（00000000）$_b$ 　=（0）$_d$ -128-2 　（10000000）$_b$ 　+（11111110）$_b$ 　=（01111110）$_b$ 　=（126）$_d$
vqadd_s8	127 127 127 -128 30 110 -110 -128
vec0	12, 12, 12, -12, 10, 10, -10, -12
vec1	12, 12, 10, -12, 20, 100, -100, -2
vadd_s8	24 24 22 -24 30 110 -110 -14
vqadd_s8	24 24 22 -24 30 110 -110 -14

mod=h 表示折半计算，如内联函数 int8x8_t vhadd_s8（int8x8_t N, int8x8_t M），用于实现 N 和 M 两个向量的对应元素相加，然后右移一位（除以 2）。其完整的测试结果见表 8-6。

表 8-6 mod=h 测试结果

操作数	执行结果
vec0	127, 127, 127, −128, 10, 10, −10, −128
vec1	127, 127, 10, −128, 20, 100, −100, −2
vadd_s8	−2 −2 −119 0 −56 −56 −56 126
vhadd_s8	127 127 68 −128 15 55 −55 −65
vec0	12, 12, 12, −12, 10, 10, −10, −12
vec1	12, 12, 10, −12, 20, 100, −100, −2
vadd_s8	24 24 22 −24 30 110 −110 −14
vhadd_s8	12 12 11 −12 15 55 −55 −7

mod=r 表示舍入计算，如内联函数 int8x8_t vrhadd_s8（int8x8_t N, int8x8_t M），用于实现 N 和 M 两个向量的对应元素相加，然后右移一位（除以 2），最终需要将得到的结果做 rounding（舍入）运算。其完整的测试结果见表 8-7。

表 8-7 mod=r 测试结果

操作数	执行结果
vec0	127, 127, 127, −128, 10, 10, −10, −128
vec1	127, 127, 10, −128, 20, 100, −100, −2
vhadd_s8	127 127 68 −128 15 55 −55 −65
vrhadd_s8	127 127 69 −128 15 55 −55 −65
vec0	12, 12, 12, −12, 10, 10, −10, −12
vec1	12, 12, 10, −12, 20, 100, −100, −2
vhadd_s8	12 12 11 −12 15 55 −55 −7
vrhadd_s8	12 12 11 −12 15 55 −55 −7

mod=d 表示加倍计算，如内联函数 int32x4_t vqdmull_s16（int16x4_t N, int16x4_t M），用于实现 N 和 M 两个向量的对应元素相乘，再乘以 2，如果结果溢出，就做饱和处理，其完整测试代码如下：

```c
#include <stdio.h>
#include <stdlib.h>
#include <arm_neon.h>
#include <math.h>

int main()
{
    int16_t vector_0[8]={1, 2, 32767, -32768, 1, 0, 7, 8};
    int16_t vector_1[8]={11, 12, 32767, -32768, 2, 16, 17, 18};
    int32_t result[8];

    int16x4_t vec0 = vld1_s16(vector_0);
    int16x4_t vec1 = vld1_s16(vector_1);

    int32x4_t vec_r = vqdmull_s16(vec0, vec1);

    vst1q_s32(result, vec_r);
    // 打印结果
    for(int i = 0; i < 8; i++){
        printf("%d", result[i]);
    }
    printf(" \n");
}
```

测试结果：

22 48 2147352578 2147483647 22 -1493953296 -1493752456 -1365370164

mod=p 表示成对（pairwise）计算，如内联函数 int8x8_t vpadd_s8（int8x8_t N, int8x8_t M），用于实现 N 和 M 两个向量中每个向量的相邻的一对元素相加，将得到的结果作为输出。

又比如内联函数 int8x8_t vpmin_s8（int8x8_t N, int8x8_t M），用于实现 N 和 M 两个向量中每个向量的相邻的一对元素的求最小值，得到的结果作为输出；同理 vpmax_s8 求最大值。

完整的测试代码如下：

```c
#include <stdio.h>
#include <stdlib.h>
#include <arm_neon.h>
```

```
int main()
{
    int8_t vector_0[8]={12, 12, 12, -12, 10, 10, -10, -12};
    int8_t vector_1[8]={12, 12, 10, -12, 20, 100, -100, -2};
    int8_t result[8];

    int8x8_t vec0 = vld1_s8(vector_0);
    int8x8_t vec1 = vld1_s8(vector_1);

    int8x8_t vec_r = vpadd_s8(vec0, vec1);
    //int8x8_t vec_r = vpmin_s8(vec0, vec1);
    //int8x8_t vec_r = vpmax_s8(vec0, vec1);
    vst1_s8(result, vec_r);

    //打印结果
    for(int i = 0; i < 8; i++){
        printf(" %d", result[i]);
    }
    printf("\n");
}
```

完整的测试结果见表 8-8。

表 8-8 mod=p 测试结果

操作数	执行结果
vec0	12, 12, 12, -12, 10, 10, -10, -12
vec1	12, 12, 10, -12, 20, 100, -100, -2
vpadd_s8	24 0 20 -22 24 -2 120 -102
vpmin_s8	12 -12 10 -12 12 -12 20 -100
vpmax_s8	12 12 10 -10 12 10 100 -2

（2）shape 后缀修饰符表示指令类型，选项有 l（Long）、n（Narrow）、w（Wide）和 q。
shape=l 表示长指令，如内联函数 int64x2_t vaddl_type（int32x2_t N, int32x2_t M）；
shape=n 表示窄指令，如内联函数 int32x4_t vaddl_s16（int16x4_t N, int16x4_t M）；
shape=w 表示宽指令，如内联函数 int64x2_t vaddw_s32（int32x2_t N, int32x2_t M）；
shape=q 表示指令操作 Q 寄存器；反之 D 寄存器。

（3）flags 表示指令标志，选项有 high、low、n 和 lane

当 flags=high 或 low 表示对向量的高半部分或低半部分操作，实现从 Q 寄存器访问 D 寄存器，如内联函数 Result_t vget_high_type（Vector_t N）和 Result_t vget_low_type（Vector_t N）分别返回输入 128 bit 向量的高半部分或低半部分，得到一个 64 bit 向量，其元素数量是输入向量的一半。完整测试代码如下：

```c
#include <stdio.h>
#include <arm_neon.h>

int main()
{
    int8_t vector_0[16] = {1, 2, 3, 4, 5, 6, 7, 8, 11, 12, 13, 14, 15, 16, 17, 18};
    int8_t result[8];

    int8x16_t vec0 = vld1q_s8(vector_0);
    int8x8_t vec_r = vget_high_s8(vec0);
    vst1_s8(result, vec_r);

    // 打印结果
    for(int i = 0; i < 8; i++){
        printf(" %d", result[i]);
    }
    printf(" \n");
}
```

代码运行结果为 11 12 13 14 15 16 17 18。

测试代码中 int8x8_t vec_r = vget_high_s8（vec0）改为 int8x8_t vec_r = vget_low_s8（vec0），代码运行结果为 1 2 3 4 5 6 7 8。

当 flags=n 表示有标量参与向量的计算，如内联函数 int32x4_t vmulq_n_s32（int32x4_t N，int32_t M），用于实现标量 M 乘以向量 N。示例代码如下：

```c
#include <stdio.h>
#include <arm_neon.h>

int main()
{
    int32_t vector_0[8] = {1, 2, 3, 4, 5, 6, 7, 8};
```

```
    int32_t result[8];
    int32_t M = 100;

    int32x4_t vec0 = vld1q_s32(vector_0);
    int32x4_t vec_r = vmulq_n_s32(vec0, M);
    vst1q_s32(result, vec_r);

    // 打印结果
    for(int i = 0; i < 8; i++){
        printf("%d", result[i]);
    }
    printf("\n");
}
```

代码运行结果为 100 200 300 400 22 -1493777168 -1493576328 -1359467828。

当 flags=lane 表示向量中某个通道参与向量的计算，如内联函数 float32_t vgetq_lane_f32(float32x4_t N, int n)，用于实现读取向量 N 中第 n 个通道的值。示例代码如下：

```
#include <stdio.h>
#include <arm_neon.h>

int main()
{
    float vector_0[8] = {1., 2., 3., 4., 5., 6., 7., 8.};
    //int n = 3;

    float32x4_t vec0 = vld1q_f32(vector_0);
    float r = vgetq_lane_f32(vec0, 3);

    // 打印结果
    printf("%f", r);
    printf("\n");
}
```

代码运行结果为 4.000000

编写代码需要注意：vgetq_lane_f32 内联函数中第二个参数需要是常数，如果传入的是变量，编译会报以下错误：

In function 'float32_t vgetq_lane_f32(float32x4_t, int)',

inlined from'int main（）'at vmulq_n_type-neon.cpp:28:29:
/opt/atk-dlrv1126-toolchain/prebuilts/gcc/linux-x86/arm/gcc-arm-8.3-2019.03-x86_64-arm-linux-gnueabihf/lib/gcc/arm-linux-gnueabihf/8.3.0/include/arm_neon.h:6267:50: error: argument 2 must be a constant immediate
　　return（float32_t）__builtin_neon_vget_lanev4sf（__a, __b）;

8.4　Neon 常用内联函数介绍

Neon 指令按照功能可分为以下：

① 类型转换指令（Intrinsics type conversion）

② 加载存储指令（Load and store）

③ 算术运算指令（Arithmetic）

④ 数据处理指令（Data processing）

⑤ 向量乘法指令（Multiply）

⑥ 逻辑和比较运算指令（Logical and compare）

包括与、或、非、等于、大于、小于等运算。

⑦ 浮点指令（Floating-point）

⑧ 移位指令（Shift）

⑨ 置换指令（Permutation）

⑩ 其他指令（Miscellaneous）[①]

8.4.1　类型转换指令

类型转换指令（Intrinsics type conversion）用于实现数据类型的转换。

1. vreinterpret

vreinterpret 指令用于对数据的强制类型转换，其内联函数定义如下：

Result_t vreinterpret_DSTtype_SRCtype（Vector1_t N）;

Result_t vreinterpretq_DSTtype_SRCtype（Vector1_t N）;

其中符号 SRCtype、DSTtype、Vector1_t、Result_t 的数据类型选项如表 8-9 所示：

---

① 多写和测试验证 Neon 代码，并且思考如何用 Neon 去优化算法。慢慢就找到感觉，能掌握如何使用指令实现高性能计算。

表 8-9　vreinterpret 指令数据类型选项

	SRCtype 或 DSTtype	Vector1_t 或 Result_t
D0—D31 寄存器数据宽度为 64 bit	s8	int8x8_t
	s16	int16x4_t
	s32	int32x2_t
	s64	int64x1_t
	u8	uint8x8_t
	u16	uint16x4_t
	u32	uint32x2_t
	u64	uint64x1_t
	f16	float16x4_t
	f32	float32x2_t
	p8	poly8x8_t
	p16	poly16x4_t
Q0—Q15 寄存器数据宽度为 128 bit	s8	int8x16_t
	s16	int16x8_t
	s32	int32x4_t
	s64	int64x2_t
	u8	uint8x16_t
	u16	uint16x8_t
	u32	uint32x4_t
	u64	uint64x2_t
	f16	float16x8_t
	f32	float32x4_t
	p8	poly8x16_t
	p16	poly16x8_t
数据类型定义：<type> <size>x<number_of_lanes>x<length_of_array>_t		

如内联函数 uint32x2_t vreinterpret_u32_s32（int32x2_t N）实现将有符号的 32 位整数（s32）转换为无符号的 32 位整数（u32）；

如内联函数 uint32x4_t vreinterpret_u32_f32（float32x4_t N）实现将有符号的 32 位浮点数（f32）转换为无符号的 32 位整数（u32）。

（2）vcombine

vcombine 指令用于将两个 64 bit 向量合并成一个 128 bit 向量，当然输出向量的元素数量是每个输入向量元素数量的两倍。其内联函数定义如下：

Result_t vcombine_type（Vector1_t N，Vector2_t M）；

具体的参数选项如表 8-10 所示。

表 8-10 内联函数参数选项

Result_t	type	Vector1_t	Vector2_t
int8x16_t	s8	int8x8_t	int8x8_t
int16x8_t	s16	int16x4_t	int16x4_t
int32x4_t	s32	int32x2_t	int32x2_t
int64x2_t	s64	int64x1_t	int64x1_t
uint8x16_t	u8	uint8x8_t	uint8x8_t
uint16x8_t	u16	uint16x4_t	uint16x4_t
uint32x4_t	u32	uint32x2_t	uint32x2_t
uint64x2_t	u64	uint64x1_t	uint64x1_t
float16x8_t	f16	float16x4_t	float16x4_t
float32x4_t	f32	float32x2_t	float32x2_t
poly8x16_t	p8	poly8x8_t	poly8x8_t
poly16x8_t	p16	poly16x4_t	poly16x4_t

8.4.2 加载存储指令

加载指令（Load）实现从内存中加载数据到 Neon 寄存器中。存储指令（Store）实现存储 Neon 寄存器里的数据到内存中。Load/Store 指令操作如图 8-9 所示。

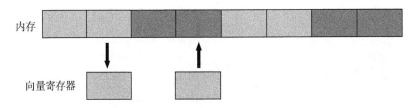

图 8-9 Load/Store 指令操作示意图

（1）vld1 和 vst1

vld1 指令用于从内存加载数据到 1 个向量，其内联函数定义如下：

Result_t vld1_type（Scalar_t* N）；

Result_t vld1q_type（Scalar_t* N）；

以上表达式中后缀不带 q，表示对应 64 bit 的向量 D 寄存器；带有 q 后缀，表示对应 128 bit 的向量 Q 寄存器。下文介绍的其他指令中后缀 q 同理。

其中符号 Result_t、type、Scalar_t 的数据类型选项如表 8-11 所示：

表 8-11 vld1 指令数据类型选项

D_0—D_{31} 寄存器 64 bit		
Result_t	type	Scalar_t
int8x8_t	s8	int8_t
int16x4_t	s16	int16_t
int32x2_t	s32	int32_t
int64x1_t	s64	int64_t
uint8x8_t	u8	uint8_t
uint16x4_t	u16	uint16_t
uint32x2_t	u32	uint32_t
uint64x1_t	u64	uint64_t
float16x4_t	f16	float16_t
float32x2_t	f32	float32_t
poly8x8_t	p8	poly8_t
poly16x4_t	p16	poly16_t
Q_0—Q_{15} 寄存器 128 bit		
Result_t	type	Scalar_t
int8x16_t	s8	int8_t
int16x8_t	s16	int16_t
int32x4_t	s32	int32_t
int64x2_t	s64	int64_t
uint8x16_t	u8	uint8_t
uint16x8_t	u16	uint16_t
uint32x4_t	u32	uint32_t
uint64x2_t	u64	uint64_t
float16x8_t	f16	float16_t
float32x4_t	f32	float32_t
poly8x16_t	p8	poly8_t
poly16x8_t	p16	poly16_t

如内联函数 uint32x4_t vld1q_f32（uint32_t const *ptr）实现以 ptr 为起始地址加载 4 个 float32 类型的数据到 Neon 向量寄存器里，其完整的示例代码如下。

```
#include <stdio.h>
#include <stdlib.h>
#include <arm_neon.h> ①
int main（）
{
    float src［12］= {0.1, 0.2, 0.3, 0.4, 0.5, 0.6, 0.7, 0.8, 0.9, 0.10, 0.11, 0.12};
    float dst［12］= {0, 0, 0, 0, 0, 0, 0, 0, 0, 0, 0, 0};
    float32x4_t vec0 = vld1q_f32（&src［1］）;
    vst1q_f32（dst, vec0）;
    for（int i = 0; i < 12; i++）{
        printf（" %f", dst［i］）;
    }
    printf（" \n"）;
}
```

代码分析：

函数 float32x4_t vec0 = vld1q_f32（&src［1］）实现一次加载以 src［1］地址起始的 4 个 float32 类型的数据到向量 vec0 中。函数 vst1q_f32（dst, vec0）实现将向量 vec0 中 4 个 float32 类型的数据赋值给以 dst 为起始地址的 4 个 float32 类型的数据。

编译代码：

（base）user@user: /neon$ arm-linux-gnueabihf-g++ test_neon.cpp -g -o test_neon

（base）user@user: /neon$ adb push test_neon /

代码运行结果：

［root@ATK-DLRV1126:/］# ./test_neon

0.200000 0.300000 0.400000 0.500000 0.000000 0.000000 0.000000 0.000000 0.000000 0.000000 0.000000 0.000000

vst1 指令实现存储 1 个向量到内存，其内联函数定义如下：

void vst1_type（Scalar_t* N, Vector_t M）;

void vst1q_type（Scalar_t* N, Vector_t M）;

---

① 头文件路径：/opt/atk-dlrv1126-toolchain/prebuilts/gcc/linux-x86/arm/gcc-arm-8.3-2019.03-x86_64-arm-linux-gnueabihf/lib/gcc/arm-linux-gnueabihf/8.3.0/include/arm_neon.h

其中符号 Result_t、type、Scalar_t 的数据类型选项如表 8-12 所示：

表 8-12 vstl 指令数据类型选项

$D_0 - D_{31}$ 寄存器 64 bit		
type	Scalar_t	Vector_t
s8	int8_t	int8x8_t
s16	int16_t	int16x4_t
s32	int32_t	int32x2_t
s64	int64_t	int64x1_t
u8	uint8_t	uint8x8_t
u16	uint16_t	uint16x4_t
u32	uint32_t	uint32x2_t
u64	uint64_t	uint64x1_t
f16	float16_t	float16x4_t
f32	float32_t	float32x2_t
p8	poly8_t	poly8x8_t
p16	poly16_t	poly16x4_t
$Q_0 - Q_{15}$ 寄存器 128 bit		
type	Scalar_t	Vector_t
s8	int8_t	int8x16_t
s16	int16_t	int16x8_t
s32	int32_t	int32x4_t
s64	int64_t	int64x2_t
u8	uint8_t	uint8x16_t
u16	uint16_t	uint16x8_t
u32	uint32_t	uint32x4_t
u64	uint64_t	uint64x2_t
f16	float16_t	float16x8_t
f32	float32_t	float32x4_t
p8	poly8_t	poly8x16_t
p16	poly16_t	poly16x8_t

以具体的内联函数 vst1q_u32（dst，vect）为例，其完整测试代码如下：

```cpp
#include <arm_neon.h>
#include <stdio.h>
#include <stdlib.h>

#include<iostream>
using namespace std;

#define LENGTH 10000
int main()
{
    int sum = 0;
    uint32_t array[LENGTH];
    for(uint i=0; i<LENGTH; i++){
        array[i]=6;
    }
    for(int i=0; i<LENGTH; i++)
        std::cout << array[i]<<" , " ; //<< std::endl;
    cout <<" \n" << endl;
    uint32x4_t in1, in2;

    uint32_t dst[LENGTH];

    //double t_1 = static_cast<double>(getTickCount());
    clock_t startTime = clock();
    for(uint i=0; i<LENGTH/4; ++i)
    {
        uint * temp =(array+4*i);
        in1 = vld1q_u32(temp);
        in2 = vld1q_u32(temp);
        in1 = vaddq_u32(in1, in2);
        vst1q_u32(dst+4*i, in1);
        //sum = vaddvq_u32(in1);
    }
    //t_1 =((double)getTickCount()-t_1)/getTickFrequency();
```

```
clock_t endTime = clock ( );
double totalTime =( double )( endTime-startTime )/ CLOCKS_PER_SEC; // 秒
cout <<" time cost:" << totalTime <<" s" << endl;

for（int i=0；i<LENGTH； i++ )
  std::cout << dst [ i ]<<" ， " ; //<< std::endl;
cout <<" \n" << endl;

return 0;
}
```

代码分析：

函数 in1 = vld1q_u32（temp）实现加载 temp 地址起始的 4 个 uint32_t 类型的数据到 in1，其中 in1 的类型为 uint32x4_t。函数 vst1q_u32（dst+4*i，in1）实现将 in1 中 4 个 uint32_t 类型的数据赋值给起始地址为 dst+4*i 空间中。

代码运行结果：

基于 V853（ARM Cortex-A7 CPU core @ 1 GHz）平台：

time cost: 0.000865292 s

基于 RV1126（四核 32 位 ARM Cortex-A7 @ 1.5 GHz）平台：

time cost: 0.000281 s

（2）vld2 和 vst2

vld2 指令用于从内存加载数据到 2 个向量，其内联函数定义如下：

Result_t vld2_type（Scalar_t* N）；

Result_t vld2q_type（Scalar_t* N）；

其中符号 Result_t、type、Scalar_t 的数据类型选项如表 8-13 所示：

表 8-13 vld2 指令数据类型选项

D_0—D_{31} 寄存器 64 bit		
Result_t	type	Scalar_t
int8x8x2_t	s8	int8_t
int16x4x2_t	s16	int16_t
int32x2x2_t	s32	int32_t
int64x1x2_t	s64	int64_t
uint8x8x2_t	u8	uint8_t
uint16x4x2_t	u16	uint16_t

D_0—D_{31} 寄存器 64 bit		
Result_t	type	Scalar_t
uint32x2x2_t	u32	uint32_t
uint64x1x2_t	u64	uint64_t
float16x4x2_t	f16	float16_t
float32x2x2_t	f32	float32_t
poly8x8x2_t	p8	poly8_t
poly16x4x2_t	p16	poly16_t
Q_0—Q_{15} 寄存器 128 bit		
Result_t	type	Scalar_t
int8x16x2_t	s8	int8_t
int16x8x2_t	s16	int16_t
int32x4x2_t	s32	int32_t
uint8x16x2_t	u8	uint8_t
uint16x8x2_t	u16	uint16_t
uint32x4x2_t	u32	uint32_t
float16x8x2_t	f16	float16_t
float32x4x2_t	f32	float32_t
poly8x16x2_t	p8	poly8_t
poly16x8x2_t	p16	poly16_t

vst2 指令实现存储 2 个向量到内存，其内联函数定义如下：

void vst2_type（Scalar_t* N，Vector_t M）;

void vst2q_type（Scalar_t* N，Vector_t M）;

以具体的内联函数 vld2q_f32 和 vst2q_f32 为例，其完整测试代码如下：

```
#include <stdio.h>
#include <stdlib.h>
#include <arm_neon.h>

int main（）
{
  float src［12］= {0.1，0.2，0.3，0.4，0.5，0.6，0.7，0.8，0.9，0.10，0.11，0.12};
```

```
float dst[12]={0, 0, 0, 0, 0, 0, 0, 0, 0, 0, 0, 0};
float dst_0[12]={0, 0, 0, 0, 0, 0, 0, 0, 0, 0, 0, 0};
float dst_1[12]={0, 0, 0, 0, 0, 0, 0, 0, 0, 0, 0, 0};
float32x4x2_t vec0 = vld2q_f32(src);

vst1q_f32(dst_0, vec0.val[0]); //float32x4_t
vst1q_f32(dst_1, vec0.val[1]);
for(int i = 0; i < 12; i++){
 printf("%f", dst_0[i]);
}
printf("\n");
for(int i = 0; i < 12; i++){
 printf("%f", dst_1[i]);
}
printf("\n");

vst2q_f32(dst, vec0);
for(int i = 0; i < 12; i++){
 printf("%f", dst[i]);
}
printf("\n");
}
```

代码运行结果：

[root@ATK-DLRV1126:/] # ./test_neon

0.100000 0.300000 0.500000 0.700000 0.000000 0.000000 0.000000 0.000000 0.000000 0.000000 0.000000 0.000000

0.200000 0.400000 0.600000 0.800000 0.000000 0.000000 0.000000 0.000000 0.000000 0.000000 0.000000 0.000000

0.100000 0.200000 0.300000 0.400000 0.500000 0.600000 0.700000 0.800000 0.000000 0.000000 0.000000 0.000000

（3）vld3 和 vst3

vld3 指令用于从内存加载数据到 3 个向量，其内联函数定义如下：

Result_t vld3_type(Scalar_t* N);

Result_t vld3q_type(Scalar_t* N);

vst3 指令实现存储 3 个向量到内存，其内联函数定义如下：

void vst3_type（Scalar_t* N，Vector_t M）;

void vst3q_type（Scalar_t* N，Vector_t M）;

通过 vld3 和 vst3 指令可以实现对数据的解交织（De-interleaving）和交织（Interleaving）操作，如图 8-10 所示。

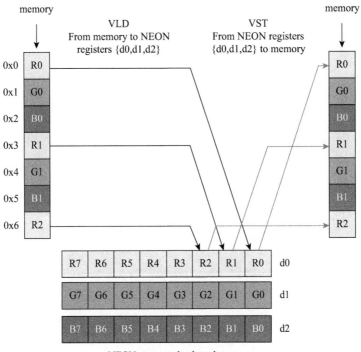

图 8-10　解交织和交织操作[①]

以具体的内联函数 vld3q_f32 和 vst3q_f32 为例，其完整测试代码如下：

```
#include <stdio.h>
#include <stdlib.h>
#include <arm_neon.h>
#include <math.h>

int main（）
{
    float src［12］= {0.1, 0.2, 0.3, 0.4, 0.5, 0.6, 0.7, 0.8, 0.9, 0.10, 0.11, 0.12};
    float dst［12］= {0, 0, 0, 0, 0, 0, 0, 0, 0, 0, 0, 0};
```

---

① 图来源：NEON Programmer's Guide Version: 1.0，Figure 6-4

```
        float dst_0[12]={0, 0, 0, 0, 0, 0, 0, 0, 0, 0, 0, 0};
        float dst_1[12]={0, 0, 0, 0, 0, 0, 0, 0, 0, 0, 0, 0};
        float dst_2[12]={0, 0, 0, 0, 0, 0, 0, 0, 0, 0, 0, 0};
        float32x4x3_t vec0 = vld3q_f32(src);

    vst1q_f32(dst_0, vec0.val[0]);  //float32x4_t
    vst1q_f32(dst_1, vec0.val[1]);
    vst1q_f32(dst_2, vec0.val[2]);
    for(int i = 0; i < 12; i++){
        printf("%f", dst_0[i]);
    }
    printf("\n");
    for(int i = 0; i < 12; i++){
        printf("%f", dst_1[i]);
    }
    printf("\n");
    for(int i = 0; i < 12; i++){
        printf("%f", dst_2[i]);
    }
    printf("\n");

    vst3q_f32(dst, vec0);
    for(int i = 0; i < 12; i++){
        printf("%f", dst[i]);
    }
    printf("\n");
}
```

代码运行结果：

[root@ATK-DLRV1126:/] # ./test_neon

　0.100000 0.400000 0.700000 0.100000 0.000000 0.000000 0.000000 0.000000 0.000000 0.000000 0.000000 0.000000

　0.200000 0.500000 0.800000 0.110000 0.000000 0.000000 0.000000 0.000000 0.000000 0.000000 0.000000 0.000000

　0.300000 0.600000 0.900000 0.120000 0.000000 0.000000 0.000000 0.000000 0.000000

0.000000 0.000000 0.000000

  0.100000 0.200000 0.300000 0.400000 0.500000 0.600000 0.700000 0.800000 0.900000

0.100000 0.110000 0.120000

（4）vld4 和 vst4

vld4 指令用于从内存加载数据到 4 个向量，其内联函数定义如下：

Result_t vld4_type（Scalar_t* N）;

Result_t vld4q_type（Scalar_t* N）;

vst4 指令实现存储 4 个向量到内存，其内联函数定义如下：

void vst4_type（Scalar_t* N, Vector_t M）;

void vst4q_type（Scalar_t* N, Vector_t M）;

以具体的内联函数 vld4q_f32 和 vst4q_f32 为例，其完整测试代码如下：

```c
#include <stdio.h>
#include <stdlib.h>
#include <arm_neon.h>
#include <math.h>

int main()
{
    float src[16] = {0.1, 0.2, 0.3, 0.4, 0.5, 0.6, 0.7, 0.8, 0.9, 0.10, 0.11, 0.12, 0.13, 0.14, 0.15, 0.16};
    float dst[16] = {0, 0, 0, 0, 0, 0, 0, 0, 0, 0, 0, 0, 0, 0, 0, 0};
    float dst_0[12] = {0, 0, 0, 0, 0, 0, 0, 0, 0, 0, 0, 0};
    float dst_1[12] = {0, 0, 0, 0, 0, 0, 0, 0, 0, 0, 0, 0};
    float dst_2[12] = {0, 0, 0, 0, 0, 0, 0, 0, 0, 0, 0, 0};
    float dst_3[12] = {0, 0, 0, 0, 0, 0, 0, 0, 0, 0, 0, 0};
    float32x4x4_t vec0 = vld4q_f32(src);

    vst1q_f32(dst_0, vec0.val[0]); // float32x4_t
    vst1q_f32(dst_1, vec0.val[1]);
    vst1q_f32(dst_2, vec0.val[2]);
    vst1q_f32(dst_3, vec0.val[3]);

    for(int i = 0; i < 12; i++){
        printf("%f", dst_0[i]);
    }
```

```
        printf ( " \n" );
        for ( int i = 0; i < 12; i++){
          printf ( " %f", dst_1 [i] );
        }
        printf ( " \n" );
        for ( int i = 0; i < 12; i++){
          printf ( " %f", dst_2 [i] );
        }
        printf ( " \n" );
        for ( int i = 0; i < 12; i++){
          printf ( " %f", dst_3 [i] );
        }
        printf ( " \n" );

        vst4q_f32 ( dst, vec0 );
        for ( int i = 0; i < 16; i++){
          printf ( " %f", dst [i] );
        }
        printf ( " \n" );
}
```

代码运行结果：

[root@ATK-DLRV1126:/] # ./test_neon

　0.100000 0.500000 0.900000 0.130000 0.000000 0.000000 0.000000 0.000000 0.000000 0.000000 0.000000 0.000000

　0.200000 0.600000 0.100000 0.140000 0.000000 0.000000 0.000000 0.000000 0.000000 0.000000 0.000000 0.000000

　0.300000 0.700000 0.110000 0.150000 0.000000 0.000000 0.000000 0.000000 0.000000 0.000000 0.000000 0.000000

　0.400000 0.800000 0.120000 0.160000 0.000000 0.000000 0.000000 0.000000 0.000000 0.000000 0.000000 0.000000

　0.100000 0.200000 0.300000 0.400000 0.500000 0.600000 0.700000 0.800000 0.900000 0.100000 0.110000 0.120000

8.4.3 算术运算指令

算术运算指令（Arithmetic）实现对整数和浮点数向量进行加减运算。文中我们以加法运算为例介绍。加法的内联函数定义如下：

Result_t vadd_type（Vector1_t N, Vector2_t M）;

Result_t vaddq_type（Vector1_t N, Vector2_t M）;

其中符号 Result_t、type、Vector1_t、Vector2_t 的数据类型选项如表 8-14 所示：

表 8-14 算术运算指令数据类型选项

D_0—D_{31} 寄存器 64 bit			
Result_t	type	Vector1_t	Vector2_t
int8x8_t	s8	int8x8_t	int8x8_t
int16x4_t	s16	int16x4_t	int16x4_t
int32x2_t	s32	int32x2_t	int32x2_t
int64x1_t	s64	int64x1_t	int64x1_t
uint8x8_t	u8	uint8x8_t	uint8x8_t
uint16x4_t	u16	uint16x4_t	uint16x4_t
uint32x2_t	u32	uint32x2_t	uint32x2_t
uint64x1_t	u64	uint64x1_t	uint64x1_t
float32x2_t	f32	float32x2_t	float32x2_t
Q_0—Q_{15} 寄存器 128 bit			
Result_t	type	Vector1_t	Vector2_t
int8x16_t	s8	int8x16_t	int8x16_t
int16x8_t	s16	int16x8_t	int16x8_t
int32x4_t	s32	int32x4_t	int32x4_t
int64x2_t	s64	int64x2_t	int64x2_t
uint8x16_t	u8	uint8x16_t	uint8x16_t
uint16x8_t	u16	uint16x8_t	uint16x8_t
uint32x4_t	u32	uint32x4_t	uint32x4_t
uint64x2_t	u64	uint64x2_t	uint64x2_t
float32x4_t	f32	float32x4_t	float32x4_t

以内联函数 vadd_u8 为例，完整测试代码如下：

示例代码 8.4.3.1 vadd_u8 测试代码

```c
#include <stdio.h>
#include <stdlib.h>
#include <arm_neon.h>
#include <math.h>

int main()
{
    uint8_t vector_0[8]={1, 2, 3, 4, 5, 6, 7, 8};
    uint8_t vector_1[8]={11, 12, 13, 14, 15, 16, 17, 18};
    uint8_t result[8];

    uint8x8_t vec0 = vld1_u8(vector_0);
    // 从数组 vector_0 中加载 8 个 uint8 类型元素存到寄存器 vec0 中
    uint8x8_t vec1 = vld1_u8(vector_1);
    uint8x8_t vec_r = vadd_u8(vec0, vec1);
    vst1_u8(result, vec_r);
    // 打印结果
    for(int i = 0; i < 8; i++){
        printf("%d", result[i]);
    }
    printf("\n");
}
```

代码运行结果：

[root@ATK-DLRV1126:/] # ./test_neon
12 14 16 18 20 22 24 26

以内联函数 vadd_s8 为例，完整测试代码如下：

```c
#include <stdio.h>
#include <stdlib.h>
#include <arm_neon.h>
#include <math.h>

int main()
{
    int8_t vector_0[8]={1, 2, 3, 4, 5, 6, 7, 8};
```

```
    int8_t vector_1[8]={11, 12, 13, 14, 15, 16, 17, 18};
    int8_t result[8];

    int8x8_t vec0 = vld1_s8(vector_0);
    // 从数组 vector_0 中 load 8 个元素存到寄存器 vec0 中
    int8x8_t vec1 = vld1_s8(vector_1);
    int8x8_t vec_r = vadd_s8(vec0, vec1);
    vst1_s8(result, vec_r);
    // 打印结果
    for(int i = 0; i < 8; i++){
        printf("%d", result[i]);
    }
    printf("\n");
}
```
代码运行结果：

[root@ATK-DLRV1126:/]# ./test_neon
12 14 16 18 20 22 24 26

8.4.4 数据处理指令

数据处理指令（Data processing）对通用的数据进行处理操作。

（1）vabs

vabs 指令实现对向量中每个 8、16 或 32 位元素的绝对值求解，其内联函数定义如下：

Result_t vabs_type(Vector_t N);

Result_t vabsq_type(Vector_t N);

具体的参数选项如表 8-15 所示。

表 8-15 vabs 指令具体参数选项

vabs 参数选项		
Result_t	type	Vector_t
int8x8_t	s8	int8x8_t
int16x4_t	s16	int16x4_t
int32x2_t	s32	int32x2_t
float32x2_t	f32	float32x2_t

续表

vabs 参数选项		
Result_t	type	Vector_t
int8x16_t	s8	int8x16_t
int16x8_t	s16	int16x8_t
int32x4_t	s32	int32x4_t
float32x4_t	f32	float32x4_t

（2）vneg

vneg 指令对向量中每个 8、16 或 32 位元素取反，其内联函数定义如下：

Result_t vneg_type（Vector_t N）;

Result_t vnegq_type（Vector_t N）;

具体的参数选项如表 8-16 所示。

表 8-16　vneg 指令参数选项

vneg 参数选项		
Result_t	type	Vector_t
int8x8_t	s8	int8x8_t
int16x4_t	s16	int16x4_t
int32x2_t	s32	int32x2_t
float32x2_t	f32	float32x2_t
vnegq 参数选项		
Result_t	type	Vector_t
int8x16_t	s8	int8x16_t
int16x8_t	s16	int16x8_t
int32x4_t	s32	int32x4_t
float32x4_t	f32	float32x4_t

（3）vmax 和 vmin

vmax 和 vmin 指令对两个向量中对应位置的 8、16 或 32 位元素依次比较大小，返回最大或最小值，其内联函数定义如下：

Result_t vmax_type（Vector1_t N, Vector2_t M）;

Result_t vmaxq_type（Vector1_t N, Vector2_t M）;

具体的参数选项如表 8-17 所示。

表 8-17 vmax 指令参数选项

vmax 参数选项			
Result_t	type	Vector1_t	Vector2_t
int8x8_t	s8	int8x8_t	int8x8_t
int16x4_t	s16	int16x4_t	int16x4_t
int32x2_t	s32	int32x2_t	int32x2_t
uint8x8_t	u8	uint8x8_t	uint8x8_t
uint16x4_t	u16	uint16x4_t	uint16x4_t
uint32x2_t	u32	uint32x2_t	uint32x2_t
float32x2_t	f32	float32x2_t	float32x2_t
vmaxq 参数选项			
Result_t	type	Vector1_t	Vector2_t
int8x16_t	s8	int8x16_t	int8x16_t
int16x8_t	s16	int16x8_t	int16x8_t
int32x4_t	s32	int32x4_t	int32x4_t
uint8x16_t	u8	uint8x16_t	uint8x16_t
uint16x8_t	u16	uint16x8_t	uint16x8_t
uint32x4_t	u32	uint32x4_t	uint32x4_t
float32x4_t	f32	float32x4_t	float32x4_t

（4）vrecpe

vrecpe 指令用于求向量中每个元素的近似倒数，其内联函数定义如下：

Result_t vrecpe_type（Vector_t N）；

Result_t vrecpeq_type（Vector_t N）；

其中 type 可选为 u32 或 f32。

（5）vrecps

vrecps 指令用于计算向量中每个元素的倒数，其内联函数定义如下：

float32x2_t vrecps_f32（float32x2_t N，float32x2_t M）；

float32x4_t vrecpsq_f32（float32x4_t N，float32x4_t M）；

（6）vrsqrte

vrsqrte 指令用于求向量中每个元素的平方根，其内联函数定义如下：

Result_t vrsqrte_type（Vector_t N）；

Result_t vrsqrteq_type（Vector_t N）；

其中 type 可选为 u32 或 f32。

（7）vcnt

vcnt 指令用于统计向量中每个元素中 bit 为 1 的个数，其内联函数定义如下：

Result_t vcnt_type（Vector_t N）；

Result_t vcntq_type（Vector_t N）；

8.4.5 向量乘法指令

向量乘法指令（Multiply）用于实现整型或浮点型向量的乘法运算。

（1）vmul

vmul 指令用于实现两个向量中的对应元素相乘，也叫点乘，其内联函数定义如下：

Result_t vmul_type（Vector1_t N, Vector2_t M）；

Result_t vmulq_type（Vector1_t N, Vector2_t M）；

（2）vmla

vmla（Vector Multiply-Add）指令实现将第二个和第三个输入向量中的对应元素点乘，然后将乘积加到第一个输入向量中的对应元素上，其内联函数定义如下：

Result_t vmla_type（Vector1_t N, Vector2_t M, Vector3_t P）；

Result_t vmlaq_type（Vector1_t N, Vector2_t M, Vector3_t P）；

8.4.6 逻辑和比较运算指令

逻辑运算指令（Logical）用于实现与、或、非等运算。

比较运算指令（Compare）用于实现等于、大于、小于等运算。

8.4.7 浮点指令

浮点指令（Floating-point）用于实现单精度浮点数、半精度浮点数、32 位整型数（Fixed-point、Integer）之间的转换操作。

（1）vcvt_n_f32

vcvt_n_f32 指令用于实现定点数转浮点数，其内联函数定义[1]如下：

Result_t vcvt_n_f32_type（Vector_t N, int n）；

Result_t vcvtq_n_f32_type（Vector_t N, int n）；

其中符号 Result_t、type、Vector_t 的数据类型选项如表 8-18 所示：

---

[1] Vector Convert

表 8-18　vcvt_n_f32 指令数据类型选项

\multicolumn{4}{c}{D_0—D_{31} 寄存器 64 bit}
Result_t	type	Vector_t	int range
float32x2_t	s32	int32x2_t	1—32
float32x2_t	u32	uint32x2_t	1—32
\multicolumn{4}{c}{Q_0—Q_{15} 寄存器 128 bit}
Result_t	type	Vector_t	int range
float32x4_t	s32	int32x4_t	1—32
float32x4_t	u32	uint32x4_t	1—32

以具体的内联函数 vcvt_n_f32_s32 为例，其完整的测试代码如下：

Code 8-1：vcvt_n_f32_s32-neon.cpp

```cpp
#include <arm_neon.h>
#include <stdio.h>

int main()
{
  int src[12]={1, 2, 3, 4, 5, 6, 7, 8, 9, 10, 11, 12};
  int32x2_t vec0 = vld1_s32(src);
  // vld1_s32 一次加载 src 地址起始的 2 个 int32 浮点数放到向量寄存器 vec0 中
  for(int i = 0; i < 12; i++)
  {
   printf("%d", src[i]);
  }
  printf("\n\n");
  //*************************
  float32x2_t vec1 = vcvt_n_f32_s32(vec0, 1);
  float dst1[12]={0, 0, 0, 0, 0, 0, 0, 0, 0, 0, 0, 0};
  vst1_f32(dst1, vec1);
  for(int i = 0; i < 12; i++)
  {
   printf("%f", dst1[i]);
  }
  printf("\n");
```

```c
//***************************
float32x2_t vec2 = vcvt_n_f32_s32(vec0, 2);
float dst2[12] = {0, 0, 0, 0, 0, 0, 0, 0, 0, 0, 0, 0};
vst1_f32(dst2, vec2);
for(int i = 0; i < 12; i++)
{
 printf(" %f", dst2[i]);
}
printf(" \n");
//***************************
float32x2_t vec3 = vcvt_n_f32_s32(vec0, 3);
float dst3[12] = {0, 0, 0, 0, 0, 0, 0, 0, 0, 0, 0, 0};
vst1_f32(dst3, vec3);
for(int i = 0; i < 12; i++)
{
 printf(" %f", dst3[i]);
}
printf(" \n");
//***************************
float32x2_t vec4 = vcvt_n_f32_s32(vec0, 4);
float dst4[12] = {0, 0, 0, 0, 0, 0, 0, 0, 0, 0, 0, 0};
vst1_f32(dst4, vec4);
for(int i = 0; i < 12; i++)
{
 printf(" %f", dst4[i]);
}
printf(" \n");

return 0;
}
```

代码运行结果：

[root@ATK-DLRV1126:/] # ./test_neon
1 2 3 4 5 6 7 8 9 10 11 12
0.500000 1.000000 0.000000 0.000000 0.000000 0.000000 0.000000 0.000000 0.000000

0.000000 0.000000 0.000000

  0.250000 0.500000 0.000000 0.000000 0.000000 0.000000 0.000000 0.000000 0.000000 0.000000 0.000000 0.000000

  0.125000 0.250000 0.000000 0.000000 0.000000 0.000000 0.000000 0.000000 0.000000 0.000000 0.000000 0.000000

  0.062500 0.125000 0.000000 0.000000 0.000000 0.000000 0.000000 0.000000 0.000000 0.000000 0.000000 0.000000

（2）vcvt_n

vcvt_n 实现浮点数转定点数，其内联函数定义如下：

Result_t vcvt_n_type_f32（Vector_t N, int n）;

Result_t vcvtq_n_type_f32（Vector_t N, int n）;

其中符号 Result_t、type、Vector_t 的数据类型选项如表 8-19 所示：

表 8-19 vcvt_n 数据类型选项

D_0—D_{31} 寄存器 64 bit			
Result_t	type	Vector_t	int range
int32x2_t	s32	float32x2_t	1—32
uint32x2_t	u32	float32x2_t	1—32
Q_0—Q_{15} 寄存器 128 bit			
Result_t	type	Vector_t	int range
int32x4_t	s32	float32x4_t	1—32
uint32x4_t	u32	float32x4_t	1—32

以内联函数 vcvt_n_s32_f32 为例，其完整测试代码如下：

```
#include <arm_neon.h>
#include <stdio.h>

int main（）
{
    float src［12］={1, 2, 3, 4, 5, 6, 7, 8, 9, 10, 11, 12};
    float32x2_t vec0 = vld1_f32（src）;
    // vld1_f32 一次加载 src 地址起始的 2 个 float32 浮点数放到向量寄存器 vec0 中
    for（int i = 0; i < 12; i++）
    {
```

```c
    printf("%f", src[i]);
}
printf("\n\n");
//***************************
int32x2_t vec1 = vcvt_n_s32_f32(vec0, 1);
int dst1[12] = {0, 0, 0, 0, 0, 0, 0, 0, 0, 0, 0, 0};
vst1_s32(dst1, vec1);
for(int i = 0; i < 12; i++)
{
    printf("%d", dst1[i]);
}
printf("\n");
//***************************
int32x2_t vec2 = vcvt_n_s32_f32(vec0, 2);
int dst2[12] = {0, 0, 0, 0, 0, 0, 0, 0, 0, 0, 0, 0};
vst1_s32(dst2, vec2);
for(int i = 0; i < 12; i++)
{
    printf("%d", dst2[i]);
}
printf("\n");
//***************************
int32x2_t vec3 = vcvt_n_s32_f32(vec0, 3);
int dst3[12] = {0, 0, 0, 0, 0, 0, 0, 0, 0, 0, 0, 0};
vst1_s32(dst3, vec3);
for(int i = 0; i < 12; i++)
{
    printf("%d", dst3[i]);
}
printf("\n");
//***************************
int32x2_t vec4 = vcvt_n_s32_f32(vec0, 4);
int dst4[12] = {0, 0, 0, 0, 0, 0, 0, 0, 0, 0, 0, 0};
vst1_s32(dst4, vec4);
```

```
    for ( int i = 0; i < 12; i++ )
    {
     printf ( "%d", dst4 [i] );
    }
    printf ( "\n" );

    return 0;
}
```

代码运行结果：

[root@ATK-DLRV1126:/] # ./test_neon

1.000000 2.000000 3.000000 4.000000 5.000000 6.000000 7.000000 8.000000 9.000000 10.000000 11.000000 12.000000

2 4 0 0 0 0 0 0 0 0 0 0
4 8 0 0 0 0 0 0 0 0 0 0
8 16 0 0 0 0 0 0 0 0 0 0
16 32 0 0 0 0 0 0 0 0 0 0

（3）vcvt_f32

vcvt_f32 实现定点数转浮点数，其内联函数定义如下：

Result_t vcvt_f32_type (Vector_t N);

Result_t vcvtq_f32_type (Vector_t N);

其中符号 Result_t、type、Vector_t 的数据类型选项如表 8-20 所示：

表 8-20　vcvt_f32 数据类型选项

D_0—D_{31} 寄存器 64 bit		
Result_t	type	Vector_t
float32x2_t	s32	int32x2_t
float32x2_t	u32	uint32x2_t
Q_0—Q_{15} 寄存器 128 bit		
Result_t	type	Vector_t
float32x4_t	s32	int32x4_t
float32x4_t	u32	uint32x4_t

以具体的内联函数 vcvt_f32_s32 为例,其完整的测试代码如下:

```c
#include <arm_neon.h>
#include <stdio.h>

int main()
{
  int src[12]={1, 2, 3, 4, 5, 6, 7, 8, 9, 10, 11, 12};
  int32x2_t vec0 = vld1_s32(src);
  // vld1_s32 一次加载 src 地址起始的 2 个 int32 整型数据放到向量寄存器 vec0 中
  for(int i = 0; i < 12; i++)
  {
    printf("%d", src[i]);
  }
  printf("\n\n");

  float32x2_t vec1 = vcvt_f32_s32(vec0);
  float dst[12]={0, 0, 0, 0, 0, 0, 0, 0, 0, 0, 0, 0};
  vst1_f32(dst, vec1);
  for(int i = 0; i < 12; i++)
  {
    printf("%f", dst[i]);
  }
  printf("\n\n");
  return 0;
}
```

代码分析:

函数 float32x2_t vec1 = vcvt_f32_s32(vec0) 实现 2 个 int32 整型数转为 float32 浮点数。

代码运行结果:

[root@ATK-DLRV1126:/] # ./test_neon

1 2 3 4 5 6 7 8 9 10 11 12

1.000000 2.000000 0.000000 0.000000 0.000000 0.000000 0.000000 0.000000 0.000000 0.000000 0.000000 0.000000

8.4.8 移位指令

移位指令（Shift）实现向量位移操作。

8.4.9 置换指令

置换指令（Permutation）实现对向量的顺序重排列操作，不修改数据的值。

（1）vrev

vrev（Reverse）为反转元素指令，包括 vrev16、vrev32、vrev64。

vrev16 指令将向量中的每个半字（16 bit）内的 8 位元素顺序反转，并将结果存储到对应的目标向量中，如内联函数 int8x8_t vrev16_s8（int8x8_t N），实现示意如图 8-11 所示。

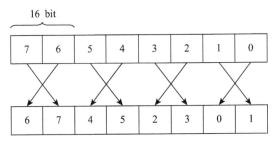

图 8-11 vrev16 指令反转示意图

vrev32 指令将向量中的每个字（32 bit）内的 8 位或 16 位元素的顺序反转，并将结果存储到对应的目标向量中。图 8-12 中，图（a）为内联函数 int8x8_t vrev32_s8（int8x8_t N）的实现示意，图（b）为内联函数 int16x4_t vrev32_s16（int16x4_t N）的实现示意。

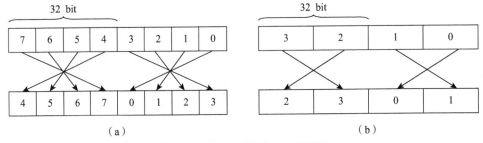

图 8-12 vrev32 指令反转示意图

vrev64 指令将向量中的每个双字（64 bit）内的 8 位、16 位或 32 位的元素顺序反转，并将结果存储到对应的目标向量中，实现示意如图 8-13 所示。

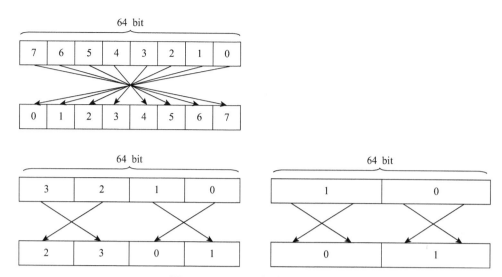

图 8-13 vrev64 指令反转示意图

（2）vtrn

vtrn（Transpose）为转置指令，可实现 2×2 小矩阵的转置操作，如图 8-14 所示。

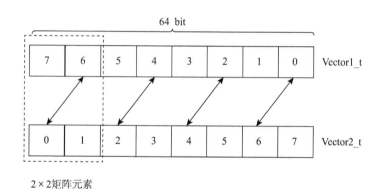

2×2 矩阵元素

图 8-14 内联函数 int8x8x2_t vtrn_s8（int8x8_t N, int8x8_t M）实现 2×2 矩阵元素转置

vtrn 内联函数定义为 Result_t vtrn_type（Vector1_t N, Vector2_t M），实现将 Vector1 中奇数索引（索引从 0 开始）的元素与 Vector2 中偶数索引的元素进行交换。具体的内联函数 int8x8x2_t vtrn_s8（int8x8_t N, int8x8_t M）测试代码如下：

```
#include <stdio.h>
#include <stdlib.h>
#include <arm_neon.h>

int main（）
{
    int8_t vector_0［8］={1, 2, 3, 4, 5, 6, 7, 8};
```

```
    int8_t vector_1[8]={9, 10, 11, 12, 13, 14, 15, 16};
    int8_t result[8];

    int8x8_t vec0 = vld1_s8(vector_0);
    // 从数组 vector_0 中 load 8 个元素存到寄存器 vec0 中
    int8x8_t vec1 = vld1_s8(vector_1);
    int8x8x2_t vec_r = vtrn_s8(vec0, vec1);
    vst1_s8(result, vec_r.val[0]);
    for(int i=0; i<8; i++){
        printf("%d", result[i]);
    }
    printf("\n");

    vst1_s8(result, vec_r.val[1]);
    for(int i=0; i<8; i++){
        printf("%d", result[i]);
    }
    printf("\n");
}
```

代码运行结果：

1 9 3 11 5 13 7 15
2 10 4 12 6 14 8 16

（3）vtbl

vtbl（Vector Table Lookup）为查表指令。其内联函数定义如下：

Result_t vtbl1_type（Vector1_t N, Vector2_t M）;

其中 M 为索引，根据索引在表 N 中查找对应位置的元素，最终输出到新的向量中。如果索引超出范围，则返回 0。表 N 使用一个 D 寄存器。该内联函数的参数选项如表 8-21 所示。

表 8-21 vtbl1 内联函数参数选项

Result_t	type	Vector1_t	Vector2_t
int8x8_t	s8	int8x8_t	int8x8_t
uint8x8_t	u8	uint8x8_t	uint8x8_t
poly8x8_t	p8	poly8x8_t	poly8x8_t

以具体的内联函数 vtbl1_s8 为例，其完整的测试代码如下：

```
#include <stdio.h>
#include <stdlib.h>
#include <arm_neon.h>

int main()
{
    int8_t vector_0[8]={1, 2, 3, 4, 5, 6, 7, 8};
    int8_t vector_1[8]={7, 6, 5, 4, 3, 2, 0, -1};
    int8_t result[8];

    int8x8_t vec0 = vld1_s8(vector_0);
    // 从数组 vector_0 中 load 8 个元素存到寄存器 vec0 中
    int8x8_t vec1 = vld1_s8(vector_1);

    int8x8_t vec_r = vtbl1_s8(vec0, vec1);

    vst1_s8(result, vec_r);
    for(int i = 0; i < 8; i++){
        printf("%d", result[i]);
    }
    printf("\n");
}
```

代码运行结果：

8 7 6 5 4 3 1 0

内联函数 Result_t vtbl2_type(Vector1_t N, Vector2_t M)，其中表 N 使用 2 个 D 寄存器，具体的参数选项如表 8-22 所示。

表 8-22 vtbl2 内联函数参数选项

Result_t	type	Vector1_t	Vector2_t
int8x8_t	s8	int8x8x2_t	int8x8_t
uint8x8_t	u8	uint8x8x2_t	uint8x8_t
poly8x8_t	p8	poly8x8x2_t	poly8x8_t

以具体的内联函数 vtbl2_s8 为例，其完整的测试代码如下：

```
#include <stdio.h>
```

```c
#include <stdlib.h>
#include <arm_neon.h>

int main()
{
    int8_t vector_0[8]={11, 12, 13, 14, 15, 16, 17, 18};
    int8_t vector_1[8]={19, 20, 21, 22, 23, 24, 25, 26};
    int8_t vector_2[8]={7, 6, 5, 4, 12, 15, 0, -1};
    int8_t result[8];

    int8x8x2_t vec0;
    vec0.val[0]=vld1_s8(vector_0);
    vec0.val[1]=vld1_s8(vector_1);
    int8x8_t vec1=vld1_s8(vector_2);

    int8x8_t vec_r=vtbl2_s8(vec0, vec1);

    vst1_s8(result, vec_r);
    for(int i=0; i<8; i++){
        printf("%d", result[i]);
    }
    printf("\n");
}
```

代码运行结果：

18 17 16 15 23 26 11 0

内联函数 Result_t vtbl3_type（Vector1_t N，Vector2_t M），其中表 N 使用 3 个 D 寄存器，具体的参数选项如表 8-23 所示。

表 8-23 vtbl3 指令参数选项

Result_t	type	Vector1_t	Vector2_t
int8x8_t	s8	int8x8x3_t	int8x8_t
uint8x8_t	u8	uint8x8x3_t	uint8x8_t
poly8x8_t	p8	poly8x8x3_t	poly8x8_t

以具体的内联函数 vtbl3_s8 为例，其完整的测试代码如下：

```c
#include <stdio.h>
```

```c
#include <stdlib.h>
#include <arm_neon.h>

int main()
{
    int8_t vector_0[8] = {11, 0, 13, 14, 15, 2, 17, -1};
    int8_t vector_1[8] = {19, 20, 21, 0, 23, 24, 25, 26};
    int8_t vector_2[8] = {27, 1, 2, 22, 6, 24, -1, 26};
    int8_t vector_index[8] = {7, 6, 5, 4, 12, 15, 0, -1};
    int8_t result[8];

    int8x8x3_t vec_table;
    vec_table.val[0] = vld1_s8(vector_0);
    vec_table.val[1] = vld1_s8(vector_1);
    vec_table.val[2] = vld1_s8(vector_2);

    int8x8_t vec_index = vld1_s8(vector_index);

    int8x8_t vec_r = vtbl3_s8(vec_table, vec_index);

    vst1_s8(result, vec_r);
    for(int i = 0; i < 8; i++){
        printf("%d", result[i]);
    }
    printf("\n");
}
```

代码运行结果：

-1 17 2 15 23 26 11 0

内联函数 Result_t vtbl4_type（Vector1_t N, Vector2_t M），其中表 N 使用 4 个 D 寄存器，具体的参数选项如表 8-24 所示。

表 8-24 vtbl4 指令参数选项

Result_t	type	Vector1_t	Vector2_t
int8x8_t	s8	int8x8x4_t	int8x8_t
uint8x8_t	u8	uint8x8x4_t	uint8x8_t
poly8x8_t	p8	poly8x8x4_t	poly8x8_t

以具体的内联函数 vtbl4_s8 为例,其完整的测试代码如下:

```c
#include <stdio.h>
#include <stdlib.h>
#include <arm_neon.h>

int main()
{
    int8_t vector_0[8]={11, 0, 13, 14, 15, 2, 17, -1};
    int8_t vector_1[8]={19, 20, 21, 0, 23, 24, 25, 26};
    int8_t vector_2[8]={27, 1, 2, 22, 6, 24, -1, 26};
    int8_t vector_3[8]={-3, 1, -2, 22, 6, 5, -1, -9};
    int8_t vector_index[8]={7, 25, 26, 4, 12, 15, 0, -1};
    int8_t result[8];

    int8x8x4_t vec_table;
    vec_table.val[0]=vld1_s8(vector_0);
    vec_table.val[1]=vld1_s8(vector_1);
    vec_table.val[2]=vld1_s8(vector_2);
    vec_table.val[3]=vld1_s8(vector_3);

    int8x8_t vec_index=vld1_s8(vector_index);

    int8x8_t vec_r=vtbl4_s8(vec_table, vec_index);

    vst1_s8(result, vec_r);
    for(int i=0; i<8; i++){
        printf(" %d", result[i]);
    }
    printf("\n");
}
```

代码运行结果:
-1 1 2 15 23 26 11 0

(4) vzip 和 vuzp

vzip(Vector Zip)为压缩指令,实现两个向量的 8、16 或 32 位元素交错组合(Interleave),具体就是先按列读后按行写。vzip 不同于 vst2,前者的结果保存在寄存器中,后者的结果保存在内存中,如图 8-15 所示。

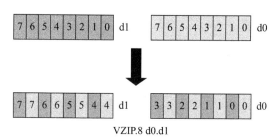

图 8-15　vzip 操作 [1]

vzip 内联函数定义如下：

Result_t vzip_type（Vector1_t N，Vector2_t M），具体的参数选项如表 8-25 所示。

表 8-25　vzip 指令参数选项

Result_t	type	Vector1_t	Vector2_t
int8x8x2_t	s8	int8x8_t	int8x8_t
int16x4x2_t	s16	int16x4_t	int16x4_t
int32x2x2_t	s32	int32x2_t	int32x2_t
uint8x8x2_t	u8	uint8x8_t	uint8x8_t
uint16x4x2_t	u16	uint16x4_t	uint16x4_t
uint32x2x2_t	u32	uint32x2_t	uint32x2_t
poly8x8x2_t	p8	poly8x8_t	poly8x8_t
poly16x4x2_t	p16	poly16x4_t	poly16x4_t
float32x2x2_t	f32	float32x2_t	float32x2_t

vzipq 的内联函数定义如下：

Result_t vzipq_type（Vector1_t N，Vector2_t M），具体的参数选项如表 8-26 所示。

表 8-26　vzipq 指令参数选项

Result_t	type	Vector1_t	Vector2_t
int8x16x2_t	s8	int8x16_t	int8x16_t
int16x8x2_t	s16	int16x8_t	int16x8_t
int32x4x2_t	s32	int32x4_t	int32x4_t
uint8x16x2_t	u8	uint8x16_t	uint8x16_t
uint16x8x2_t	u16	uint16x8_t	uint16x8_t

---

[1] 图来源：https://community.arm.com/arm-community-blogs/b/architectures-and-processors-blog/posts/coding-for-neon---part-5-rearranging-vectors

续表

Result_t	type	Vector1_t	Vector2_t
uint32x4x2_t	u32	uint32x4_t	uint32x4_t
poly8x16x2_t	p8	poly8x16_t	poly8x16_t
poly16x8x2_t	p16	poly16x8_t	poly16x8_t
float32x4x2_t	f32	float32x4_t	float32x4_t

vuzp（Vector Unzip）为解压指令，实现两个向量的8、16或32位元素解交错组合（de-interleave），具体就是先按行读后按列写。vuzp不同于vld2，前者实现从寄存器到寄存器的数据操作，后者实现从内存加载数据到寄存器，如图8-16所示。

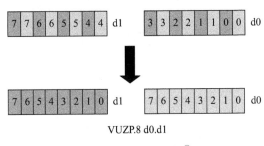

图8-16 vuzp操作①

vuzp的内联函数定义如下：
Result_t vuzp_type（Vector1_t N, Vector2_t M），具体的参数选项如表8-27所示。

表8-27 vuzp指令参数选项

Result_t	type	Vector1_t	Vector2_t
int8x8x2_t	s8	int8x8_t	int8x8_t
int16x4x2_t	s16	int16x4_t	int16x4_t
int32x2x2_t	s32	int32x2_t	int32x2_t
uint8x8x2_t	u8	uint8x8_t	uint8x8_t
uint16x4x2_t	u16	uint16x4_t	uint16x4_t
uint32x2x2_t	u32	uint32x2_t	uint32x2_t
poly8x8x2_t	p8	poly8x8_t	poly8x8_t
poly16x4x2_t	p16	poly16x4_t	poly16x4_t
float32x2x2_t	f32	float32x2_t	float32x2_t

① 图来源：https://community.arm.com/arm-community-blogs/b/architectures-and-processors-blog/posts/coding-for-neon---part-5-rearranging-vectors vuzpq 的内联函数定义如下：

Result_t vuzpq_type（Vector1_t N，Vector2_t M），具体的参数选项如表 8-28 所示。

表 8-28 vuzpq 指令参数选项

Result_t	type	Vector1_t	Vector2_t
int8x16x2_t	s8	int8x16_t	int8x16_t
int16x8x2_t	s16	int16x8_t	int16x8_t
int32x4x2_t	s32	int32x4_t	int32x4_t
uint8x16x2_t	u8	uint8x16_t	uint8x16_t
uint16x8x2_t	u16	uint16x8_t	uint16x8_t
uint32x4x2_t	u32	uint32x4_t	uint32x4_t
poly8x16x2_t	p8	poly8x16_t	poly8x16_t
poly16x8x2_t	p16	poly16x8_t	poly16x8_t
float32x4x2_t	f32	float32x4_t	float32x4_t

8.4.10 其他指令

其他指令（Miscellaneous）包括一些杂项的指令，比如 vcreate、vdup_n、vdup_lane、vmove_n 等。

（1）vcreate

vcreate 指令实现创建一个 64bit 的向量，内联函数定义如下：

Result_t vcreate_type（Scalar_t N）；

其中符号 Result_t、type、Scalar_t 的数据类型选项如表 8-29 所示：

表 8-29 vcreate 数据类型选项

Result_t	type	Scalar_t
int8x8_t	s8	uint64_t
int16x4_t	s16	uint64_t
int32x2_t	s32	uint64_t
int64x1_t	s64	uint64_t
uint8x8_t	u8	uint64_t
uint16x4_t	u16	uint64_t
uint32x2_t	u32	uint64_t
uint64x1_t	u64	uint64_t
float16x4_t	f16	uint64_t

续表

Result_t	type	Scalar_t
float32x2_t	f32	uint64_t
poly8x8_t	p8	uint64_t
poly16x4_t	p16	uint64_t

以内联函数 vcreate_u16 为例,其完整测试代码如下:

```
#include <stdio.h>
#include <stdlib.h>
#include <arm_neon.h>

int main(void)
{
    uint16x4_t vector; // define v as a vector with 4 lanes of 16-bit data
    unsigned short result[4]; // 分配4个16bit的内存块
    vector = vcreate_u16(0x000A000301020012);
    vst1_u16(result, vector); // store the vector to memory, in this case, to array result

    for(int i=0; i<4; i++)
        printf("%d", result[i]);
    printf("\n");
    unsigned char c = result[1];
    printf("%d", c);
printf("\n");
}
```

代码分析:

函数 vector = vcreate_u16(0x000A000301020012)创建一个向量,其中的值从低位到高位分别为 18 258 3 10,如图 8-17 所示。

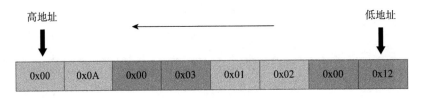

图 8-17 vcreate_u16

编译方法：
$ arm-linux-gnueabihf-g++ neon-vcreate_u8.cpp -g -o test_neon
代码运行结果：
［root@ATK-DLRV1126:/］# ./test_neon
18 258 3 10
2

（2）vdup_n

vdup_n 指令将标量复制到目标向量的每个元素中，其内联函数定义如下：

Result_t vdup_n_type（Scalar_t N）；

Result_t vdupq_n_type（Scalar_t N）；

（3）vdup_lane

vdup_lane 指令将标量复制到目标向量的每个元素中，其内联函数定义如下：

Result_t vdup_lane_type（Vector_t N, int n）；

Result_t vdupq_lane_type（Vector_t N, int n）；

（4）vmov_n

vmov_n 指令将标量复制到目标向量的每个元素中，其内联函数定义如下：

Result_t vmov_n_type（Scalar_t N）；

Result_t vmovq_n_type（Scalar_t N）；

8.5　Neon 编程优化算法实例

8.5.1　RGB 转 Gray 颜色空间

RGB 转 Gray（灰度图）是很多图像算法的组成部分，其中 RGB 转 Gray 的心理学计算公式为：

$$\text{gray} = r \times 0.299 + g \times 0.587 + b \times 0.114 \tag{8.5-1}$$

式（8.5-1）中由于采用浮点计算，往往性能没有定点高。于是可以优化为采用 16 位定点数表示和移位计算实现除法，具体见式（8.5-2）。

$$\text{gray} = \begin{cases} r \times \text{int}(0.299 \times 2^{16}) + g \times \text{int}(0.587 \times 2^{16}) + \\ b \times [2^{16} - \text{int}(0.299 \times 2^{16}) - \text{int}(0.587 \times 2^{16})] \end{cases} >> 16 \tag{8.5-2}$$

$$\text{gray} = (r \times 19595 + g \times 38469 + b \times 7472) >> 16$$

还可以进一步压缩到 8 位定点数表示，见式（8.5-3）。

$$\text{gray} = \begin{Bmatrix} r \times \text{int}(0.299 \times 2^8) + g \times \text{int}(0.587 \times 2^8) + \\ b \times [\, 2^8 - \text{int}(0.299 \times 2^8) - \text{int}(0.587 \times 2^8)\,] \end{Bmatrix} >> 8 \qquad (8.5\text{-}3)$$

$$\text{gray} = [\, r \times 76 + g \times 150 + b \times (256 - 76 - 150)\,] >> 8$$

$$\text{gray} = (r \times 76 + g \times 150 + b \times 30) >> 8$$

$$2^8 = 256$$

以上公式用 C 语言代码实现如下：

```c
for( i=0; i<n; i++ ){
    int r = *src++;
    int g = *src++;
    int b = *src++;
    int y = ( r*19595 )+( g*38469 )+( b*7472 );
    *dest++ =( y>>16 );
}
```

Neon 编程代码如下：

```c
// 读取 8 字节预设值到 64 位寄存器
uint8x8_t rfac = vdup_n_u8（76）;
uint8x8_t gfac = vdup_n_u8（150）;
uint8x8_t bfac = vdup_n_u8（30）;
n/=8;
for( i=0; i<n; i++ )
{
    uint16x8_t temp;
// 一次读取 3 个 unit8x8 到 3 个 64 位寄存器
    uint8x8x3_t rgb = vld3_u8（src）;
    uint8x8_t result;

    temp = vmull_u8（rgb.val [ 0 ], rfac）;
    temp = vmlal_u8（temp, rgb.val [ 1 ], gfac）;
    temp = vmlal_u8（temp, rgb.val [ 2 ], bfac）;
    result = vshrn_n_u16（temp, 8）;
    vst1_u8（dest, result）;
    src += 8*3;
    dest += 8;
}
```

其中内联函数 vld3q_u8（…）的实现如图 8-18 所示：

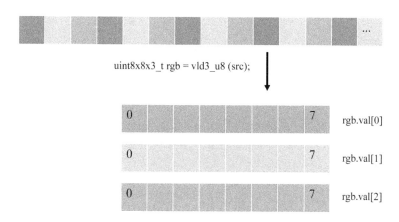

图 8-18　vld3q_u8（…）操作示意图

8.5.2　RGB 内存空间解交织和交织

vld3q_u8（）实现 RGB 内存数据排列顺序解交织（HWC->CHW 解交织），也叫打包（Packed）格式和平面（Planar）格式的转换，如图 8-19 所示。

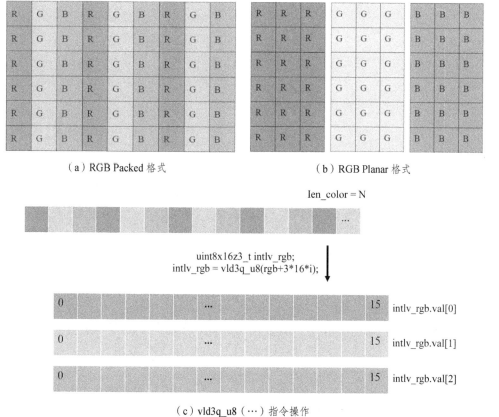

图 8-19　RGB 内存数据排列顺序解交织

完整测试代码如下：

```c
//#if __ARM_NEON

//#if __ARM_NEON
#include <arm_neon.h>
//#endif // __ARM_NEON

#include <stdlib.h>
#include <stdio.h>
#include <memory.h>
//#include <fstream>

#include <sys/time.h>

//code from https://developer.arm.com/documentation/102467/0100/Example---RGB-
//deinterleaving
void rgb_deinterleave_c ( uint8_t *r, uint8_t *g, uint8_t *b, uint8_t *rgb, int len_color )
{
    /* Take the elements of "rgb" and store the individual colors "r", "g", and "b" .*/
    for ( int i=0; i < len_color; i++ )
    {
        r [ i ]= rgb [ 3*i ];
        g [ i ]= rgb [ 3*i+1 ];
        b [ i ]= rgb [ 3*i+2 ];
    }
}

void rgb_deinterleave_neon ( uint8_t *r, uint8_t *g, uint8_t *b, uint8_t *rgb, int len_color )
{
    /*Take the elements of "rgb" and store the individual colors "r", "g", and "b" .*/
    int num8x16 = len_color / 16;
    // len_color 的值等于 W*H*3，为什么除以 16 呢，因为 Neon 寄存器的宽度为 128bit，
    // 也就是 16 个 Byte，
    // 这个向量寄存器一次可以加载 16 个字节。
    uint8x16x3_t intlv_rgb;
    for ( int i=0; i < num8x16; i++ )
```

```cpp
    {
        intlv_rgb = vld3q_u8（rgb+3*16*i）;
        // vld3q_u8 函数实现一次加载 3 个大小为 16 个字节的向量
        vst1q_u8（r+16*i, intlv_rgb.val［0］）;
        // vst1q_u8 函数实现一次存储大小为 16 个字节的向量
        vst1q_u8（g+16*i, intlv_rgb.val［1］）;
        vst1q_u8（b+16*i, intlv_rgb.val［2］）;
    }
}
//

int W = 960;
int H = 540;
struct timeval time_begin, time_end;
unsigned char *RGBRGBRGB_buf;
unsigned char *RRRGGGBBB_buf;

int main（int argc, char **argv）
{
    int pixel_size=W*H;
    int img_size=W*H*3;
    RGBRGBRGB_buf=（unsigned char *）malloc（img_size）;
    RRRGGGBBB_buf=（unsigned char *）malloc（img_size）;

    for（int i = 0; i < pixel_size; i++）
    {
        RGBRGBRGB_buf［3*i+0］=10;
        RGBRGBRGB_buf［3*i+1］=128;
        RGBRGBRGB_buf［3*i+2］=200;
    }
    //std::fstream file1（"./RGBRGBRGB_buf.dat", std::ios::binary|std::ios::out）;
    //file1.write（（char *）RGBRGBRGB_buf, img_size）;
    //file1.close（）;

    int time = 0;
    // c time cost
```

gettimeofday（&time_begin, NULL）;

　rgb_deinterleave_c（RRRGGGBBB_buf, RRRGGGBBB_buf + pixel_size, RRRGGGBBB_buf + 2*pixel_size, RGBRGBRGB_buf, pixel_size）;

gettimeofday（&time_end, NULL）;

　time =（1000000*（time_end.tv_sec-time_begin.tv_sec）+（time_end.tv_usec-time_begin.tv_usec））/1000;

printf（"c cost: %d ms\n", time）;

//std::fstream file2（"./RRRGGGBBB_buf_c.dat", std::ios::binary|std::ios::out）;
//file2.write（(char *) RRRGGGBBB_buf, img_size）;
//file2.close（）;

// neon time cost

gettimeofday（&time_begin, NULL）;

　rgb_deinterleave_neon（RRRGGGBBB_buf, RRRGGGBBB_buf + pixel_size, RRRGGGBBB_buf + 2*pixel_size, RGBRGBRGB_buf, pixel_size）;

gettimeofday（&time_end, NULL）;

　time =（1000000*（time_end.tv_sec-time_begin.tv_sec）+（time_end.tv_usec-time_begin.tv_usec））/1000;

printf（"neon cost: %d ms\n", time）;

//std::fstream file3（"./RRRGGGBBB_buf_neon.dat", std::ios::binary|std::ios::out）;
//file3.write（(char *) RRRGGGBBB_buf, img_size）;
//file3.close（）;

return 0;

}

在 RV1126 平台的运行性能评估如下：

（1）RV1126 平台 CPU 测试

编译：

$ arm-linux-gnueabihf-g++ neon_deinterleave.c -g -o test_neon

运行结果如表 8-31 所示：

表 8-31　RV1126 平台 CPU 运行性能对比

RV1126 CPU	pure c cost	Neon cost
默认编译配置	24 ms	5 ms
编译中配置 -O3 优化参数	3 ms	4 ms

(2) V853 平台 CPU (ARM Cortex-A7@1GHz, 32 KB I-cache, 32 KB D-cache, 128 KB L2 cache) 测试

编译和测试:

$ /V853/tina-v853-open/prebuilt/rootfsbuilt/arm/toolchain-sunxi-musl-gcc-830/toolchain/bin/arm-openwrt-linux-gcc -o test_neon_deinterleave test_neon_deinterleave.c

$ adb push test_neon_deinterleave home/test

运行结果如表 8-32 所示:

表 8-32 V853 平台 CPU 运行性能对比

V853 平台 CPU	pure c cost	Neon cost
默认编译配置	40 ms	4 ms
编译中配置 -O3 优化参数	8 ms	6 ms

8.5.3 矩阵乘法性能优化

关于矩阵乘法的 C 语言实现, 其代码如下[①]:

```
void matrix_multiply_c ( float32_t *A, float32_t *B, float32_t *C, uint32_t n, uint32_t m, uint32_t k )
{
    for( int i_idx=0; i_idx<n; i_idx++ )
    {
        for( int j_idx=0; j_idx<m; j_idx++ )
        {
            C [ n*j_idx + i_idx ]= 0;
            for( int k_idx=0; k_idx<k; k_idx++ )
            {
                C [ n*j_idx + i_idx ]+=
                    A [ n*k_idx + i_idx ] *B [ k*j_idx + k_idx ];
            }
        }
    }
}
```

---

① 出自: https://developer.arm.com/documentation/102467/0201/Example---matrix-multiplication

接下来我们讨论如何使用Neon来优化矩阵乘法。float32x4_t类型的Neon寄存器中的向量通道为4,每个通道数据类型为float32。我们将大尺寸矩阵分解为多个4×4小矩阵,再优化4×4小矩阵乘,矩阵乘的流程如图8-20所示。细心的读者朋友会发现和线性代数里的流程有点不同,但对编程实现来说是等效的。

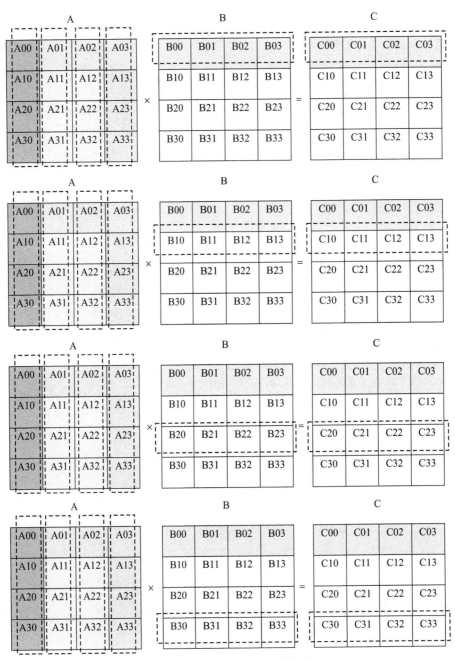

图 8-20 矩阵 A 乘矩阵 B 输出矩阵示意图

矩阵乘的计算方式如下：

C00 = B00 × A00 + B01 × A10 + B02 × A20 + B03 × A30

C01 = B00 × A01 + B01 × A11 + B02 × A21 + B03 × A31

C02 = B00 × A02 + B01 × A12 + B02 × A22 + B03 × A32

C03 = B00 × A03 + B01 × A13 + B02 × A23 + B03 × A33

公式可以变换为

C00 =（A00，A10，A20，A30）⊙（B00，B01，B02，B03）

C01 =（A01，A11，A21，A31）⊙（B00，B01，B02，B03）

C02 =（A02，A12，A22，A32）⊙（B00，B01，B02，B03）

C03 =（A03，A13，A23，A33）⊙（B00，B01，B02，B03）

其中⊙代表点积（dot product）。

A0 =（A00，A10，A20，A30）

A1 =（A01，A11，A21，A31）

A2 =（A02，A12，A22，A32）

A3 =（A03，A13，A23，A33）

B0 =（B00，B01，B02，B03）

通过 Neon 指令实现：

C0 = vfmaq_laneq_f32（C0，A0，B0，0）；

C0 = vfmaq_laneq_f32（C0，A1，B0，1）；

C0 = vfmaq_laneq_f32（C0，A2，B0，2）；

C0 = vfmaq_laneq_f32（C0，A3，B0，3）；

vfmaq_laneq_f32[①]指令实现单指令周期里一次乘加运算，只有 Armv8-A、Armv9-A 架构等支持。在 Armv7-A 架构 CPU 中可以将 vfmaq_laneq_f32 指令通过 vgetq_lanc_f32、vmulq_n_f32、vaddq_f32 指令组合实现，当然这样性能会有影响。将

C0 = vfmaq_laneq_f32（C0，A0，B0，0）；

C0 = vfmaq_laneq_f32（C0，A1，B0，1）；

C0 = vfmaq_laneq_f32（C0，A2，B0，2）；

C0 = vfmaq_laneq_f32（C0，A3，B0，3）；

修改为：

C0 = vaddq_f32（vmulq_n_f32（A0，vgetq_lane_f32（B0，0）），C0）；

C0 = vaddq_f32（vmulq_n_f32（A1，vgetq_lane_f32（B0，1）），C0）；

C0 = vaddq_f32（vmulq_n_f32（A2，vgetq_lane_f32（B0，2）），C0）；

---

① Vector Fused Multiply Accumulate，融合乘加运算

C0 = vaddq_f32（vmulq_n_f32（A3，vgetq_lane_f32（B0，3）），C0）;

其他位置的代码修改同理。

其中 vmulq_n_f32 指令实现向量浮点数乘以常数（Scalar），定义如下：

Result_t vmulq_n_type（Vector_t N，Scalar_t M）;

vaddq_f32 指令实现向量浮点数加法，定义如下：

Result_t vaddq_type（Vector1_t N，Vector2_t M）;

代码编译：

$ arm-linux-gnueabihf-g++ matrix-multiplication-neon.cpp -O3 -g -o test_neon

运行结果如表 8-33 所示：

表 8-33　运行性能对比

RV1126 CPU 矩阵大小均为 400×400	pure c cost	Neon cost
编译中配置 -O3 优化参数	1141 ms	297 ms

8.6　NEON2SSE 介绍

NEON2SSE 是 Intel 提供的用于将 NEON 内联函数转化为 Intel SIMD-SSE 内联函数的接口。通过 NEON2SSE 可实现基于 X86 平台开发 SIMD 程序，还可以实现在 X86 平台上调试 ARM 平台的 NEON 程序。NEON2SSE 使用起来简单，只需在我们的代码中添加以下头文件就可以。

```
#ifdef ARM_PLATFORM
#include <arm_neon.h>
#else
#include "NEON_2_SSE.h"
#endif
// https://github.com/intel/ARM_NEON_2_x86_SSE
```

第 9 章

基于 TFLite 的端侧模型部署和性能优化

本章主要讨论如何使用 TFLite（TensorFlow Lite）框架实现端侧模型的部署，其内容包括 TFLite 交叉编译、TFLite FlatBuffer 模型文件转换和验证、端侧模型部署和性能评估以及性能优化。本章中用于测试的模型包括 YOLOv8 和 Transformer。TFLite 方案相比较其他方案，其中一个显著优势是完全开源。

9.1　TFLite 委托

TFLite 委托（Delegate）是指在模型部署中利用设备端的加速器（例如 GPU、DSP、NPU 等）来运行 TFLite 模型的部分或全部网络。通过委托 API 去调用设备中加速器的 API，让开发更加简单。TFLite 包括转换（Converter）和解析（Interpreter）两部分，前者用于 TFLite FlatBuffer 模型文件转换，包括优化和量化，后者用于模型加载、Buffer 分配、推理和委托等，如图 9-1 所示。

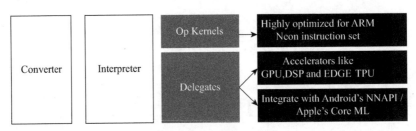

图 9-1　TFLite 委托流程[①]

9.2　TFLite 交叉编译和部署——基于 ARM 平台

我们的需求是把一个训练好的模型部署在 ARM 平台，并且性能满足需求。首先使用命令 cat /proc/cpuinfo 查看部署平台 CPU 的详细信息，包括 CPU 型号、核数、线程数、CPU 主频以及缓存信息等。以 RV1126 平台为例，进入平台终端后，运行以上命令得到的信息如图 9-2 所示，可以知道该平台具有 4 个 ARM Cortex-A7 核，且每个核均支持 Neon。

---

① 图来源：https://tensorflow.google.cn/lite/performance/delegates?hl=zh-cn；https://tensorflow.google.cn/lite/performance/implementing_delegate?hl=zh-cn

```
[root@ATK-DLRV1126:/demo]# cat /proc/cpuinfo
processor       : 0
model name      : ARMv7 Processor rev 5 (v7l)
BogoMIPS        : 121.00
Features        : half thumb fastmult vfp edsp neon vfpv3 tls vfpv4 idiva idivt vfpd32 lpae
processor       : 1
model name      : ARMv7 Processor rev 5 (v7l)
BogoMIPS        : 121.00
Features        : half thumb fastmult vfp edsp neon vfpv3 tls vfpv4 idiva idivt vfpd32 lpae
processor       : 2
model name      : ARMv7 Processor rev 5 (v7l)
BogoMIPS        : 121.00
Features        : half thumb fastmult vfp edsp neon vfpv3 tls vfpv4 idiva idivt vfpd32 lpae
processor       : 3
model name      : ARMv7 Processor rev 5 (v7l)
BogoMIPS        : 121.00
Features        : half thumb fastmult vfp edsp neon vfpv3 tls vfpv4 idiva idivt vfpd32 lpae
```

图 9-2 CPU 的详细信息（图有改动）

9.2.1 用 CMake 工具构建 TFLite

第 1 步，克隆 TensorFlow 仓库。

git clone https://github.com/tensorflow/tensorflow.git

其版本信息如下：

tensorflow$ git log

commit 63f5a65c7cd7b6241bede8d2e0082058566ea364（HEAD, tag: v2.15.1, origin/r2.15）

第 2 步，创建 CMake 构建目录。

mkdir build_arm

cd build_arm

第 3 步，使用配置运行 CMake 工具。

tensorflow/build_arm$

ARMCC_FLAGS=" -march=armv7-a -mfpu=neon-vfpv4 -funsafe-math-optimizations -mfp16-format=ieee"

ARMCC_PREFIX=/opt/atk-dlrv1126-toolchain/bin/arm-linux-gnueabihf-

cmake -DCMAKE_C_COMPILER=${ARMCC_PREFIX}gcc \

  -DCMAKE_CXX_COMPILER=${ARMCC_PREFIX}g++ \

  -DCMAKE_C_FLAGS=" ${ARMCC_FLAGS}" \

  -DCMAKE_CXX_FLAGS=" ${ARMCC_FLAGS}" \

  -DCMAKE_VERBOSE_MAKEFILE:BOOL=ON \

  -DCMAKE_SYSTEM_NAME=Linux \

  -DCMAKE_SYSTEM_PROCESSOR=armv7 \

  ../tensorflow/lite/

先说明一下以上的多个编译选项，具体内容如下：

-march=armv7-a 表示编译出的目标文件是针对 Armv7 架构。

-mfpu=neon-vfpv4 用于指定目标处理器的浮点单元（FPU）类型。其中 neon 表示使用 Neon 指令集进行加速运算，vfpv4 表示使用 VFPv4 指令集进行浮点数运算。这个选项通常用于支持 ARM 架构的处理器，以优化浮点数运算的性能。

-funsafe-math-optimizations，用于启用 GCC 编译器的不安全数学优化。当启用此选项时，编译器会尝试对数学运算进行优化，以提高程序的性能，但这可能导致某些情况下的数值精度损失。如果需要确保数值精度，建议禁用此选项。

-mfp16-format=ieee 用于指定浮点数的表示格式。mfp16 表示使用半精度浮点数（Half-precision Floating Point），ieee 表示使用 IEEE 754 标准来表示浮点数。这个选项通常用于嵌入式系统或需要节省存储空间的场景。

-DCMAKE_VERBOSE_MAKEFILE:BOOL=ON 表示 CMake 构建过程中输出详细的编译命令、链接命令等构建过程信息。

-DCMAKE_CXX_COMPILER 指定 C++ 编译器的路径，-DCMAKE_C_FLAGS 指定 C 编译器的路径。

-DCMAKE_SYSTEM_PROCESSOR=armv7 指定目标系统的处理器架构是 ARM 架构 32 位处理器。

如果在配置运行构建中出现如下错误：

/tensorflow/build_arm/xnnpack/src/qs8-qc8w-igemm/gen/qs8-qc8w-igemm-2x2c4-minmax-fp32-armsimd32.c:80:15: error: unknown type name'int16x2_t'
　　　　const int16x2_t va1c02 = __sxtb16（va1）;
…………

/tensorflow/build_arm/xnnpack/src/qs8-qc8w-igemm/gen/qs8-qc8w-igemm-2x2c4-minmax-fp32-armsimd32.c:83:24: error: expected'='，'，'，';'，'asm'or'__attribute__'before'vb0'
　　　　const int8x4_t vb0 = *（(const int8x4_t*）w）; w =（const int8_t*）w + 4;
/tensorflow/build_arm/xnnpack/src/qs8-qc8w-igemm/gen/qs8-qc8w-igemm-2x2c4-minmax-fp32-armsimd32.c:83:24: error:'vb0'undeclared（first use in this function）; did you mean'va0'?
　　　　const int8x4_t vb0 = *（(const int8x4_t*）w）; w =（const int8_t*）w + 4;
　　　　　　va0
…………

/tensorflow/build_arm/xnnpack/src/qs8-qc8w-igemm/gen/qs8-qc8w-igemm-2x2c4-minmax-fp32-armsimd32.c:83:30: error: invalid type argument of unary'*'（have'int'） const int8x4_t vb0 = *((const int8x4_t*)w); w = (const int8_t*)w + 4;

…………

/tensorflow/build_arm/xnnpack/src/qs8-qc8w-igemm/gen/qs8-qc8w-igemm-2x2c4-minmax-fp32-armsimd32.c:92:24: error: expected'=','，','；','asm'or'__attribute__'before'vb1'

const int8x4_t vb1 = *((const int8x4_t*)w); w = (const int8_t*)w + 4;

…………

/tensorflow/build_arm/xnnpack/src/qs8-qc8w-igemm/gen/qs8-qc8w-igemm-2x2c4-minmax-fp32-armsimd32.c:146:20: error:'voutput_min'undeclared(first use in this function)

/……/tensorflow/build_arm/xnnpack/src/qs8-qc8w-igemm/gen/qs8-qc8w-igemm-2x2c4-minmax-fp32-armsimd32.c:146:44: error: expected'；'before'params'

const int8x4_t voutput_min = (int8x4_t)params->fp32_armsimd32.output_min;

…………

/tensorflow/build_arm/xnnpack/src/qs8-qc8w-igemm/gen/qs8-qc8w-igemm-2x2c4-minmax-fp32-armsimd32.c:150:44: error: expected'；'before'params'

const int8x4_t voutput_max = (int8x4_t)params->fp32_armsimd32.output_max;

我们对比问题进行分析：

在错误日志中，我们可以看到int16x2_t和int8x4_t均是32bit的，而Neon支持的位宽是64bit或128bit，所以问题出在XNNPACK的编译选项配置。相关的配置如图9-3所示，其中Ruy、Eigen和XNNPACK均为用于TensorFlow Lite的加速库。

选项名称	功能	Android	Linux	macOS	Windows
TFLITE_ENABLE_RUY	启用 RUY	开	关	关	关
::matrix:::::					
::multiplication:::::					
::library:::::					
TFLITE_ENABLE_NNAPI	启用 NNAPI	开	关	不适用	不适用
::委托:::::					
TFLITE_ENABLE_GPU	启用 GPU	关	关	不适用	不适用
::委托:::::					
TFLITE_ENABLE_XNNPACK	启用 XNNPACK	开	开	开	开
::委托:::::					
TFLITE_ENABLE_MMAP	启用 MMAP	开	开	开	不适用

图 9-3 TensorFlow Lite 选项列表 [1]

---

[1] 图来源：https://tensorflow.google.cn/lite/guide/build_cmake?hl=zh-cn

Ruy 是一个矩阵乘法库，用于加速神经网络推理中的矩阵乘法运算，特别优化了小尺寸矩阵的乘法运算。TensorFlow Lite 编译可配置 Ruy 模块，用于支持浮点和 8 位整数量化矩阵。①

Eigen 是专门实现线性代数运算的高性能 C++ 模板库，其支持矩阵、向量、数值求解器等相关计算。②

XNNPACK 是一个高度优化的用于 ARM、X86、WebAssembly 和 RISC-V 平台的神经网络推理加速解决方案。XNNPACK 提供基本的优化算子，用于支持机器学习框架，比如 TensorFlow Lite、TensorFlow.js、PyTorch、ONNX Runtime 和 MediaPipe 等。③默认情况下 XNNPACK 处于启用状态。

具体解决方法如下。

方法 1：

添加编译配置 -DTFLITE_ENABLE_XNNPACK=OFF 禁用 XNNPACK，再重新运行 CMake，问题可以解决。

方法 2：

当然方法 1 没有从根本上解决问题。接下来尝试升级交叉编译器去解决，升级命令如下：④

sudo apt update

sudo apt install gcc-arm-none-eabi -y

sudo apt install gcc-arm-linux-gnueabihf -y

sudo apt install gcc-aarch64-linux-gnu -y

交叉编译器更新成功，版本信息如图 9-4 所示：

图 9-4　更新交叉编译器后的版本信息

---

① https://github.com/google/ruy.git
② 更多资料可参考：http://eigen.tuxfamily.org/，https://gitlab.com/libeigen/eigen.git
③ https://github.com/google/XNNPACK.git
④ https://learn.arm.com/install-guides/gcc/cross/

第 9 章 基于 TFLite 的端侧模型部署和性能优化

重新运行命令构建工程，完整的命令如下：

ARMCC_FLAGS="-march=armv7-a -mfpu=neon-vfpv4 -funsafe-math-optimizations -mfp16-format=ieee"

ARMCC_PREFIX=arm-linux-gnueabihf-

cmake -DCMAKE_C_COMPILER=${ARMCC_PREFIX}gcc \
　-DCMAKE_CXX_COMPILER=${ARMCC_PREFIX}g++ \
　-DCMAKE_C_FLAGS=" ${ARMCC_FLAGS}" \
　-DCMAKE_CXX_FLAGS=" ${ARMCC_FLAGS}" \
　-DCMAKE_VERBOSE_MAKEFILE:BOOL=ON \
-DTFLITE_ENABLE_XNNPACK=ON \
　-DTFLITE_ENABLE_RUY=ON \
-DCMAKE_SYSTEM_NAME=Linux \
-DCMAKE_SYSTEM_PROCESSOR=armv7 \
../tensorflow/lite/

构建过程中，如果出现不能下载的文件，可以提前下载好，直接拷贝到指定的路径，举例说明：如果地址 https://storage.googleapis.com/mirror.tensorflow.org/github.com/intel/ARM_NEON_2_x86_SSE/archive/a15b489e1222b2087007546b4912e21293ea86ff.tar.gz 在构建中下载失败，可以先下载好文件后拷贝到路径 build_arm/_deps/neon2sse-subbuild/neon2sse-populate-prefix/src 中。

但是前面的问题依然没有得到解决。需要进一步去分析定位，在下一小节重点讨论如何去解决这个问题。

9.2.2 XNNPACK 编译问题定位和解决

本小节编译 TensorFlow 源码中路径为 tensorflow/lite/examples/minimal 的测试实例。先编译为 X86 环境下可执行文件。

第 1 步，build$ cmake../tensorflow/lite/examples/minimal。

其中完整的 CMakeLists.txt 文件如图 9-5 所示。

第 2 步，build$ make -j8。

第 3 步，编译成功得到可执行文件 minimal，去进一步验证模型的推理性能，如图 9-6 所示。模型验证成功，推理时间为 1878ms，测试硬件为 X86 平台，其硬件配置为 Intel（R）Core（TM）i7-6800K CPU @ 3.40GHz 12 核。

build$ ./minimal yolov8m_float32.tflite

```cmake
cmake_minimum_required(VERSION 3.16)
project(minimal C CXX)

set(TENSORFLOW_SOURCE_DIR "" CACHE PATH
  "Directory that contains the TensorFlow project"
)
if(NOT TENSORFLOW_SOURCE_DIR)
  get_filename_component(TENSORFLOW_SOURCE_DIR
    "${CMAKE_CURRENT_LIST_DIR}/../../../../"
    ABSOLUTE
  )
endif()

add_subdirectory(
  "${TENSORFLOW_SOURCE_DIR}/tensorflow/lite"
  "${CMAKE_CURRENT_BINARY_DIR}/tensorflow-lite"
  EXCLUDE_FROM_ALL
)

set(CMAKE_CXX_STANDARD 17)
add_executable(minimal
  minimal.cc
)
target_link_libraries(minimal
  tensorflow-lite
)
```

图 9-5　CMakeLists.txt 文件

```
---------------Memory Arena Status Start---------------
Total memory usage: 34406400 bytes (32.812 MB)
- Total arena memory usage: 34406400 bytes (32.812 MB)
- Total dynamic memory usage: 0 bytes (0.000 MB)

Subgraph#0   Arena (Normal)          34406400 (100.00%)
---------------Memory Arena Status End---------------

inference time in ms: 1878
```

图 9-6　模型验证

说明 XNNPACK 支持 X86 环境，现在的问题是：为什么 XNNPACK 在 ARM 平台交叉编译不可以？

我们将 XNNPACK 回退到指定版本：

XNNPACK$ git checkout

40a672f8b29d85ac4b4718a4fbbf7249c8ef1141

然后基于 Armv7-a 交叉编译，编译配置选项如下：

ARMCC_FLAGS=" -march=armv7-a -mfpu=neon-vfpv4 -funsafe-math-optimizations -mfp16-format=ieee"

ARMCC_PREFIX=/opt/atk-dlrv1126-toolchain/bin/arm-linux-gnueabihf-
cmake -DCMAKE_C_COMPILER=${ARMCC_PREFIX}gcc \
 -DCMAKE_CXX_COMPILER=${ARMCC_PREFIX}g++ \
 -DCMAKE_C_FLAGS=" ${ARMCC_FLAGS}" \
 -DCMAKE_CXX_FLAGS=" ${ARMCC_FLAGS}" \
 -DCMAKE_VERBOSE_MAKEFILE:BOOL=ON \
 -DCMAKE_SYSTEM_NAME=Linux \
 -DCMAKE_SYSTEM_PROCESSOR=armv7 \
 -DCMAKE_BUILD_TYPE=Debug \
　..

最终编译成功。说明当前最新版本的 XNNPACK（bbbaa7352a3ea729987d3e654d37be93e8009691）支持基于 Android 系统的 Armv7（Neon）处理器，但是对于 Linux 系统却没有提供对应的脚本，而之前的版本是支持的。

经过测试验证，目前最简单的解决方案是将 TensorFlow 版本回退到 v2.7.0，具体如下：

v2.7.0/tensorflow$ git log

commit c256c071bb26e1e13b4666d1b3e229e110bc914a（HEAD, tag: v2.7.0）

v2.7.0/tensorflow/build/xnnpack$ git log

commit 694d2524757f9040e65a02c374e152a462fe57eb（HEAD）

完整的验证步骤如下：

第 1 步，修改 CMakeLists.txt 文件。

具体修改如下：

cmake_minimum_required（VERSION 3.16）

set（tools /opt/atk-dlrv1126-toolchain/bin/arm-linux-gnueabihf-）
set（CMAKE_C_COMPILER ${tools}gcc）
set（CMAKE_CXX_COMPILER ${tools}g++）
project（minimal）

第 2 步，创建 CMake 构建目录。

tensorflow$ mkdir minimal_build

tensorflow$ cd build_arm

第 3 步，使用配置运行 CMake 工具。

ARMCC_FLAGS=" -march=armv7-a -mfpu=neon-vfpv4 -funsafe-math-optimizations -mfp16-format=ieee"

```
ARMCC_PREFIX=/opt/atk-dlrv1126-toolchain/bin/arm-linux-gnueabihf-
cmake -DCMAKE_C_COMPILER=${ARMCC_PREFIX}gcc \
 -DCMAKE_CXX_COMPILER=${ARMCC_PREFIX}g++ \
 -DCMAKE_C_FLAGS=" ${ARMCC_FLAGS}" \
 -DCMAKE_CXX_FLAGS=" ${ARMCC_FLAGS}" \
 -DCMAKE_VERBOSE_MAKEFILE:BOOL=ON \
 -DTFLITE_ENABLE_XNNPACK=OFF \
 -DTFLITE_ENABLE_RUY=ON \
 -DCMAKE_SYSTEM_NAME=Linux \
 -DCMAKE_SYSTEM_PROCESSOR=armv7 \
 ../tensorflow/lite/examples/minimal
```

CMake 阶段中如果出现下载错误：

error: downloading' https://storage.googleapis.com/mirror.tensorflow.org/github.com/petewarden/OouraFFT/archive/v1.0.tar.gz' failed

可以提前下载好，然后放到指定的路径 minimal_build/_deps/fft2d-subbuild/fft2d-populate-prefix/src 中，问题解决。

第 4 步，make 编译源码。

运行命令 make -j8。

第 5 步，验证。

编译成功后，得到的可执行文件在 RV1126 平台运行验证，方法如下：

minimal_build$ adb push minimal /test

minimal_build$ adb push yolov8m_float32.tflite /test

最终测试用例运行成功。

9.3 YOLOv8 模型端侧部署——低空无人机巡检

低空无人机巡检是通过无人机载荷传感器设备实现不同场景的巡检需求，从而提高巡检效率和质量，节省维护成本。如图 9-7 所示的无人机森林巡检，它可以飞过山头，不畏悬崖峭壁，不分白天黑夜，实现日常的森林巡查，高效排查野外用火、非法砍伐、盗采、乱倒垃圾等违规行为。此外其搭载的红外热成像设备，可准确检测到森林中温度异常区域，对于森林防火发挥巨大的作用。通过无人机森林巡检设备，不仅提高巡检的质量，还可大大节约人力投入。

（a）无人机森林巡检中实现燃烧火焰检测

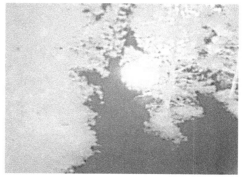
（b）热成像

图 9-7　无人机森林巡检 [①]

无人机高速公路巡检成为高速公路安全巡检的重要方式。虽然高速公路上安装了很多监控设备，但是无法做到全面覆盖，无人机可弥补其视野盲区。无人机用于监控道路异常停车、行人违规上高速、车辆拥堵等现象，具有应急响应快、灵活性高、监控范围广的特点，而且可实现 24 小时无人值守自动巡查，确保道路畅通和安全。

以上讨论的两种无人机巡检场景，我们都需要用到计算机视觉中的目标检测算法，通过该算法可实时检测到以上场景中的燃烧火焰、车辆、行人等目标。本节我们讨论基于 TFLite 实现 YOLOv8 目标检测模型在端侧的部署。

准备工作开始，首先克隆工程代码：

$ git clone https://github.com/ultralytics/ultralytics.git
ultralytics$ git log
commit 5c1277113b19e45292c01e5a47aa2bdb6ebc98d0（HEAD -> main，origin/main，origin/HEAD）

9.3.1　TFLite FlatBuffer 模型文件转换和验证

下载 yolov8m.pt 模型作为测试模型 [②]，如图 9-8 所示：

[①] 图来源：北亚利桑那大学公开的基于航拍图像的森林火情检测数据集 FLAME。
[②] 模型地址：https://github.com/ultralytics/assets/releases/download/v8.1.0/yolov8m.pt，该模型支持 80 类物体检测，包括：0: person 1: bicycle 2: car 3: motorcycle 4: airplane 5: bus 6: train 7: truck 8: boat 9: traffic light 10: fire hydrant 11: stop sign 12: parking meter 13: bench 14: bird 15: cat 16: dog 17: horse 18: sheep 19: cow 20: elephant 21: bear 22: zebra 23: giraffe 24: backpack 25: umbrella 26: handbag 27: tie 28: suitcase 29: Frisbee 30: skis 31: snowboard 32: sports ball 33: kite 34: baseball bat 35: baseball glove 36: skateboard 37: surfboard 38: tennis racket 39: bottle 40: wine glass 41: cup 42: fork 43: knife 44: spoon 45: bowl 46: banana 47: apple 48: sandwich 49: orange 50: broccoli 51: carrot 52: hot dog 53: pizza 54: donut 55: cake 56: chair 57: couch 58: potted plant 59: bed 60: dining table 61: toilet 62: tv 63: laptop 64: mouse 65: remote 66: keyboard 67: cell phone 68: microwave 69: oven 70: toaster 71: sink 72: refrigerator 73: book 74: clock 75: vase 76: scissors 77: teddy bear 78: hair drier 79: toothbrush

Model	size (pixels)	mAP^{val} 50-95	Speed CPU ONNX (ms)	Speed A100 TensorRT (ms)	params (M)	FLOPs (B)
YOLOv8n	640	37.3	80.4	0.99	3.2	8.7
YOLOv8s	640	44.9	128.4	1.20	11.2	28.6
YOLOv8m	640	50.2	234.7	1.83	25.9	78.9
YOLOv8l	640	52.9	375.2	2.39	43.7	165.2
YOLOv8x	640	53.9	479.1	3.53	68.2	257.8

图 9-8 YOLOv8 不同体积的模型 [1]

将 yolov8m.pt 模型转换为 TFLite FlatBuffer 格式的方法有两种。

第一种，使用 CL 命令，但是需要先安装 ultralytics。具体的转换命令为 yolo mode=export model=yolov8m.pt format=tflite。

第二种，使用 Python 代码，具体代码如下：

pttotflite.py

from ultralytics import YOLO

model = YOLO（'yolov8m.pt'）

results = model.export（format='tflite', int8=True）

成功运行以上代码需要的环境包版本为：onnx2tf>=1.15.4，<=1.17.5，sng4onnx>=1.0.1，onnxsim>=0.4.33，onnx_graphsurgeon>=0.3.26，tflite_support== 0.4.4，tflite_runtime-2.14.0，onnxruntime== 1.17.1，tensorflow-cpu==2.13.1[2]，torch==2.1.1+cpu。

运行命令 ultralytics$ python pttotflite.py，成功转换的部分日志如下：

ultralytics$ python pttotflite.py

Ultralytics YOLOv8.1.27 🚀 Python-3.11.5 torch-2.1.1+cpu CPU（Intel Core（TM）i7-6800K 3.40GHz）

YOLOv8m summary（fused）：218 layers, 25886080 parameters, 0 gradients

PyTorch: starting from 'yolov8m.pt' with input shape（1, 3, 640, 640）BCHW and output shape（s）（1, 84, 8400）（49.7 MB）

TensorFlow SavedModel: starting export with tensorflow 2.13.1...

---

[1] 图来源：https://github.com/ultralytics/ultralytics?tab=readme-ov-file

[2] pip install tensorflow-cpu==2.13.1 -i https://mirrors.aliyun.com/pypi/simple/

//············

saved_model output complete!

Float32 tflite output complete!

Float16 tflite output complete!

Input signature information for quantization

signature_name: serving_default

input_name.0: images shape:(1, 640, 640, 3)dtype: <dtype:'float32'>

Dynamic Range Quantization tflite output complete!

fully_quantize: 0, inference_type: 6, input_inference_type: FLOAT32, output_inference_type: FLOAT32

INT8 Quantization tflite output complete!

fully_quantize: 0, inference_type: 6, input_inference_type: INT8, output_inference_type: INT8

Full INT8 Quantization tflite output complete!

INT8 Quantization with int16 activations tflite output complete!

Full INT8 Quantization with int16 activations tflite output complete!

TensorFlow SavedModel: export success ✓ 476.7s, saved as'yolov8m_saved_model'（323.2 MB）

TensorFlow Lite: starting export with tensorflow 2.13.1...

TensorFlow Lite: export success ✓ 0.0s, saved as'yolov8m_saved_model/yolov8m_int8.tflite'（25.0 MB）

Export complete（478.9s）

Results saved to /······/ultralytics

Predict: yolo predict task=detect model=yolov8m_saved_model/yolov8m_int8.tflite imgsz=640 int8

Validate: yolo val task=detect model=yolov8m_saved_model/yolov8m_int8.tflite imgsz=640 data=coco.yaml int8

Visualize: https://netron.app

转化成功后得到的模型文件如图9-9所示，其中_float32.tflite为全精度模型，其参数为32位浮点数（float32）存储；_float16.tflite为半精度模型，其参数为16位浮点数（float16）存储；_int8.tflite为量化整型模型，其参数为8位整型数（int8）存储。

```
/ultralytics/yolov8m_saved_model$ ls -l --block-size=m
total 373M
  1M assets
  1M fingerprint.pb
  1M metadata.yaml
100M saved_model.pb
  1M variables
 50M yolov8m_float16.tflite
100M yolov8m_float32.tflite
 25M yolov8m_full_integer_quant.tflite
 26M yolov8m_int8.tflite
 25M yolov8m_integer_quant.tflite
 50M yolov8m.pt
```

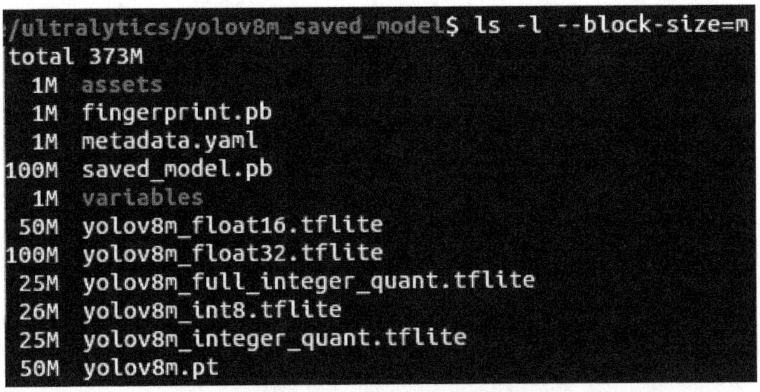

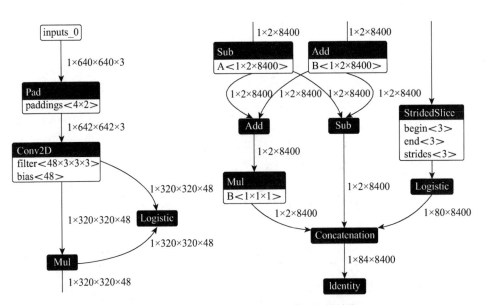

图 9-9　TFLite 模型转换得到的模型文件

yolov8m_float32.tflite 模型网络结构（部分）如图 9-10 所示，其中输入张量尺寸为（1，3，640，640），输出张量尺寸为（1，84，8400）。网络结构图使用工具 https://netron.app/ 查看。

图 9-10　yolov8m_float32.tflite 模型网络结构

下一步需要验证 TFLite 模型转换是否成功。我们在硬件平台中先验证 yolov8m_float32.tflite 模型，具体如下：

```
[root@ATK-DLRV1126:/test]# ./minimal ../demo/yolov8m_float32.tflite
=== Pre-invoke Interpreter State ===
Interpreter has 1 subgraphs.

//……
```

---------------Memory Arena Status Start---------------
---------------Memory Arena Status End---------------

说明模型可以在 ARM 平台加载和解析，但是模型功能是否有问题，需要进一步验证。

先在 PC 端验证 *.tflite 模型，检测效果如图 9-11 所示：

图 9-11　yolov8m_float32.tflite 模型在 PC 端推理效果

9.3.2　ARM 平台模型部署——C 语言版

在 ARM 平台中实现模型的部署，一般使用 C 语言。首先介绍一个 C 语言代码的调试技巧，我们使用以下代码可以把模型输入和模型输出的数据保存到 *.txt 文件中，然后在 Python 验证代码中加载这个 *.txt 文件，可以很方便地帮助我们定位到 C 语言代码的问题。

运行在端侧的代码：

```
#include <cstdio> //C++ 的标准输入输出库
//#include <stdio.h> //C 的标准输入输出库
*FILE *file;
file = fopen ( "file_outputData.txt" , "w" ); // 打开文件， "w" 表示写入模式
if ( file == NULL ) {
  printf ( "无法打开文件 \n" );
  return 0;
}
for ( int i=0; i<length; i++ )
{
  printf ( " %f, " , outputBuffer [ i ] );
```

```cpp
    char* content;
    sprintf(content, "%f", outputBuffer[i]); // 浮点转 char 类型
    fprintf(file, "%s,", content); // 将内容写入文件
}
//std::cout << std::endl;
fclose(file); // 关闭文件
```

```python
# Python
file_path = 'file_outputData.txt'
    with open(file_path, 'r', encoding='utf-8') as file:
    content = file.read()
print(content)
data = content.split(',')
data_array = np.zeros(len(data), dtype=np.float32)
for i in range(len(data)-1):
    data_array[i] = float(data_[i])
output = data_array.reshape(1, 84, 8400)
```

部署代码构建脚本如下：

```
export OpenCV_PATH=/……/opencv-4.9.0
ARMCC_FLAGS=" -march=armv7-a -mfpu=neon-vfpv4 -fopenmp -funsafe-math-optimizations -fPIC -lrt -D_GNU_SOURCE -lpthread -lm -ldl -lopencv_world -I ${OpenCV_PATH}/install/include -L ${OpenCV_PATH}/install/lib"
ARMCC_PREFIX=/opt/atk-dlrv1126-toolchain/bin/arm-linux-gnueabihf-
cmake -DCMAKE_C_COMPILER=${ARMCC_PREFIX}gcc \
  -DCMAKE_CXX_COMPILER=${ARMCC_PREFIX}g++ \
  -DCMAKE_C_FLAGS=" ${ARMCC_FLAGS}" \
  -DCMAKE_CXX_FLAGS=" ${ARMCC_FLAGS}" \
  -DCMAKE_VERBOSE_MAKEFILE:BOOL=ON \
  -DTFLITE_ENABLE_XNNPACK=ON \
  -DTFLITE_ENABLE_RUY=ON \
  -DCMAKE_SYSTEM_NAME=Linux \
  -DCMAKE_SYSTEM_PROCESSOR=armv7 \
  -DCMAKE_BUILD_TYPE=Debug \
  ../tensorflow/lite/examples/infer
```

部署工程的 CMakeLists.txt 文件的内容如下：

cmake_minimum_required（VERSION 3.16）

set（tools /opt/atk-dlrv1126-toolchain/bin/arm-linux-gnueabihf-）

set（CMAKE_C_COMPILER ${tools}gcc）

set（CMAKE_CXX_COMPILER ${tools}g++）

project（infer）#C CXX

set（TENSORFLOW_SOURCE_DIR" " CACHE PATH
 " Directory that contains the TensorFlow project"
）

if（NOT TENSORFLOW_SOURCE_DIR）
 get_filename_component（TENSORFLOW_SOURCE_DIR
  " ${CMAKE_CURRENT_LIST_DIR}/../../../../"
  ABSOLUTE
 ）
endif（）

add_subdirectory（
 " ${TENSORFLOW_SOURCE_DIR}/tensorflow/lite"
 " ${CMAKE_CURRENT_BINARY_DIR}/tensorflow-lite"
 EXCLUDE_FROM_ALL
）

sct（CMAKE_CXX_STANDARD 11）
add_executable（infer
 infer.cc
）
target_link_libraries（infer
 tensorflow-lite
）

完整 C 语言代码如下：

#include <cstdio>
//#include <stdio.h>

#include" tensorflow/lite/interpreter.h"

```cpp
#include"tensorflow/lite/kernels/register.h"
#include"tensorflow/lite/model.h"
#include"tensorflow/lite/optional_debug_tools.h"

#include<opencv2/opencv.hpp>
#include <sys/time.h>

using namespace std;
using namespace cv;

#define TFLITE_MINIMAL_CHECK(x)                              \
  if(!(x)){                                                  \
    fprintf(stderr,"Error at %s:%d\n",__FILE__,__LINE__);    \
    exit(1);                                                 \
  }

float cls_thres = 0.5;
float iou_thres = 0.5;

int img_width = 640;
int img_height = 640;
int image_channels = 3;

struct detect_obj {
    float cx, cy, w, h;        // 边界框坐标：中心点，宽，高
    float x1, y1, x2, y2;      // 边界框坐标：左上角，右下角
    float score;               // 置信度
    int label;                 // 标签
};

cv::Mat resize_padding(const cv::Mat &image, int target_width, int target_height){
    cv::Mat resized_img;
    double scale = min(1.0 * target_width / image.cols, 1.0 * target_height / image.rows);
    cv::resize(image, resized_img, cv::Size(), scale, scale, cv::INTER_LINEAR);

    // 计算需要填充的空白区域大小
```

```cpp
    int paddingX = target_width - resized_img.cols;
    int paddingY = target_height - resized_img.rows;

    // 创建填充后的图像
    cv::Mat padded_img（target_height, target_width, resized_img.type（）, cv::Scalar
（100, 100, 100））;

    // 将缩放后的图像复制到填充后的图像中心
    cv::Rect roi（（paddingX > 0 ? paddingX / 2 : 0）, （paddingY > 0 ? paddingY / 2 : 0）,
        resized_img.cols, resized_img.rows）;
    resized_img.copyTo（padded_img（roi））;

    return padded_img;
}

std::vector<uchar> mat_vector（const cv::Mat &image）
{
    std::vector<uchar> img_vector;
    if（!image.empty（））
    {
        int total_elements = image.total（）* image.channels（）;
        img_vector.resize（total_elements）;
        int index = 0;

        for（int i = 0; i < image.rows; ++i）
        {
            for（int j = 0; j < image.cols; ++j）
            {
                cv::Vec3b pixel = image.at<cv::Vec3b>（i, j）;
                img_vector［index++］= pixel［2］; // R channel
                img_vector［index++］= pixel［1］; // G channel
                img_vector［index++］= pixel［0］; // B channel
            }
        }
    }
    return img_vector;
```

```cpp
}
std::vector<float> normalize ( const std::vector<uchar> &vec )
{
    std::vector<float> normalized_vec;
    for ( float value: vec )
    {
        value /= 255.0;
        normalized_vec.push_back ( value );
    }
    return normalized_vec;
}

int main ( int argc, char* argv [ ] ){
    if( argc != 3 ){
        fprintf ( stderr, " model <tflite model>\n" );
        return 1;
    }
    const char* filename = argv [ 1 ];
    const char* testimg = argv [ 2 ]; //" test_1.jpg"

    // 记录启动时间
    std::chrono::steady_clock::time_point start, end;

    // Load model
    std::unique_ptr<tflite::FlatBufferModel> model =
        tflite::FlatBufferModel::BuildFromFile ( filename );
    TFLITE_MINIMAL_CHECK ( model != nullptr );

    // Build the interpreter with the InterpreterBuilder.
    // Note: all Interpreters should be built with the InterpreterBuilder,
    // which allocates memory for the Interpreter and does various set up
    // tasks so that the Interpreter can read the provided model.
    tflite::ops::builtin::BuiltinOpResolver resolver;
    tflite::InterpreterBuilder builder ( *model, resolver );
    std::unique_ptr<tflite::Interpreter> interpreter;
```

```cpp
builder(&interpreter);
TFLITE_MINIMAL_CHECK(interpreter != nullptr);

// Allocate tensor buffers.
TFLITE_MINIMAL_CHECK(interpreter->AllocateTensors() == kTfLiteOk);
printf("=== Pre-invoke Interpreter State ===\n");
tflite::PrintInterpreterState(interpreter.get());

// Fill input buffers
float *input_tensor = interpreter->typed_input_tensor<float>(0);

Mat input_img = imread(testimg);
cv::Mat img = resize_padding(input_img, img_width, img_height);
//imwrite("img_640_640.jpg", img);

printf("load the image ***************************\n");
if(img.empty()){
    printf("load the image error \n");
    return -1;
}

// Mat 转 vector
std::vector<uchar> img_vector = mat_vector(img);
// 归一化
std::vector<float> input_data = normalize(img_vector);

// Get the input tensor pointer
int inputTensorIndex = interpreter->inputs()[0];

// Get input buffer of type float
auto *inputBuffer = interpreter->typed_tensor<float>(inputTensorIndex);

start = std::chrono::steady_clock::now();
memcpy(inputBuffer, input_data.data(), input_data.size() * sizeof(float));
end = std::chrono::steady_clock::now();
auto test_time = std::chrono::duration_cast<std::chrono::milliseconds>(end - start).
```

```cpp
count();
    std::cout <<" Copy input_data data to the input tensor cost time in ms:" + std::to_string(test_time) << std::endl;

    start = std::chrono::steady_clock::now();
    // Inference
    interpreter->Invoke();

    // Get Output
    int outputTensorIndex = interpreter->outputs()[0];
    //printf(" outputTensorIndex: %d ****************\n", outputTensorIndex);

    // Get output buffer of type float
    float* outputBuffer = interpreter->typed_tensor<float>(outputTensorIndex);

    // copy output buffer to outputData
    unsigned long long totalElements = interpreter->tensor(outputTensorIndex)->bytes / sizeof(float);
    std::vector<float> outputData(outputBuffer, outputBuffer + totalElements);

    end = std::chrono::steady_clock::now();
    auto inference_time = std::chrono::duration_cast<std::chrono::milliseconds>(end - start).count();
    std::cout <<" Inference Time in ms:" + std::to_string(inference_time) << std::endl;

    printf(" totalElements: %d *************************\n", totalElements);

    // 获取输出矩阵维度信息
    TfLiteIntArray *output_dims = interpreter->tensor(interpreter->outputs()[0])->dims;

    const unsigned int num_anchors = output_dims->data[2]; //8400
    const unsigned int num_classes = output_dims->data[1] - 4; //80

    std::cout <<" num_anchors :" + std::to_string(num_anchors) +" , num_classes :" + std::to_string(num_classes) << std::endl;

    float *output_ptr = outputBuffer; //outputdata.data();
```

```cpp
std::vector<detect_obj> detect_results;
std::vector<cv::Rect> boxesNMS;
std::vector<float> confidences;
std::vector<int> indices;

// 遍历锚点
for( unsigned int i = 0; i < num_anchors; ++i )
{
    const float *row_ptr = output_ptr + i;

    // 类别置信度
    float max_cls_conf = 0.0;
    int best_cls_idx = 0;
    for( unsigned int c = 0; c < num_classes; ++c )
    {
        float cls_conf = row_ptr [ num_anchors*（4 + c）];
        if( cls_conf > max_cls_conf )
        {
            max_cls_conf = cls_conf;
            best_cls_idx = c;
        }
    }

    // 计算边界框坐标
    const float *offsets = row_ptr;
    float cx = offsets [ num_anchors * 0 ]* img_width;
    float cy = offsets [ num_anchors * 1 ]* img_height;
    float w = offsets [ num_anchors * 2 ]* img_width;
    float h = offsets [ num_anchors * 3 ]* img_height;

    detect_obj obj;
    float x1 =（ cx - w / 2.f）;
    float y1 =（ cy - h / 2.f）;
    float x2 =（ cx + w / 2.f）;
    float y2 =（ cy + h / 2.f）;
```

```cpp
        x1 = std::max ( 0.f, x1 );
        y1 = std::max ( 0.f, y1 );
        //x2 = std::min ( ( float ) img_width, x2 );
        //y2 = std::min ( ( float ) img_height, y2 );

        obj.label = best_cls_idx;
        obj.score = max_cls_conf;

        detect_results.push_back ( obj );
        boxesNMS.push_back ( cv::Rect ( x1, y1, w, h ) );
        confidences.push_back ( max_cls_conf );
    }

    cv::dnn::NMSBoxes ( boxesNMS, confidences, cls_thres, iou_thres, indices );

    // 输出结果
    for ( int i = 0; i < indices.size ( ); i++ )
    {
        cv::Rect rect = boxesNMS [ indices [ i ] ];
        cv::Point topleft ( ( int )( rect.x ), ( int )( rect.y ) );
        cv::Point bottomRight ( ( int )( rect.x + rect.width ), ( int )( rect.y + rect.height ) );
        // 右下角坐标
        cv::rectangle ( img, topleft, bottomRight, cv::Scalar ( detect_results [ indices [ i ] ].label, 0, 255 ), 1 );

        // 添加标签文字
        std::string label = std::to_string ( detect_results [ indices [ i ] ].label );
        std::string score = std::to_string ( detect_results [ indices [ i ] ].score );
        std::string label_score = label + " " + score;
        double fontScale = 0.5;
        int thickness = 1;
        int baseline = 0;
        cv::Size textSize = cv::getTextSize ( label_score, cv::FONT_HERSHEY_SIMPLEX, fontScale, thickness, &baseline );
        cv::Point text_position ( topleft.x, topleft.y - 10 );
```

```
        // 画一个背景矩形
        cv::rectangle（img, text_position,
                text_position + cv::Point（textSize.width, -textSize.height）, cv::Scalar
（detect_results [indices [i]] .label, 0, 255), -1）;
        // 将标签文字放在矩形上
        cv::putText（img, label_score, text_position, cv::FONT_HERSHEY_SIMPLEX,
fontScale, cv::Scalar（0, 0, 0）, thickness）;
    }
    imwrite（"out.jpg", img）;
    return 0;
}
```

测试命令：

v2.7.0/tensorflow/build$ adb push infer /demo

[root@ATK-DLRV1126:/demo] # ./infer yolov8m_int8.tflite test_1.jpg

测试结果如图 9-12 所示：

图 9-12　量化模型 yolov8m_int8.tflite 端侧检测结果

如果出现以下问题：

./infer: error while loading shared libraries: libgomp.so.1: cannot open shared object file: No such file or directory

解决方法：

将交叉编译器中的 libgomp.so 文件拷贝到测试端，具体命令为：

user@user:/opt/atk-dlrv1126-toolchain/arm-buildroot-linux-gnueabihf/sysroot/lib$ adb push libgomp.so.1 /demo

user@user:/opt/atk-dlrv1126-toolchain/arm-buildroot-linux-gnueabihf/sysroot/lib$ adb push libgomp.so /demo

9.3.3 部署性能优化

验证模型 yolov8m_float32.tflite 在 PC 机上的推理时间为 1862ms（测试环境为 X86 平台，其配置为 Intel（R）Core（TM）i7-6800K CPU @ 3.40GHz 12 核）。

再验证模型 yolov8m_float32.tflite 在 ARM 平台（RV1126@CPU 核）上的推理时间，得到表 9-1。

表 9-1 yolov8m_float32.tflite 模型的推理时间

-DTFLITE_ENABLE_XNNPACK	ON	ON	OFF	OFF
-DTFLITE_ENABLE_RUY	ON	OFF	ON	OFF
Inference Time/ms	51627	51682	84063	369408

注：测试命令 ./infer yolov8m_float32.tflite

当配置选项为：

-DTFLITE_ENABLE_XNNPACK=ON -DTFLITE_ENABLE_RUY=ON

ARM 平台（RV1126@CPU 核）测试整型量化和半精度浮点数模型结果为：

[root@ATK-DLRV1126:/demo]# ./infer yolov8m_int8.tflite

Inference Time in ms: 43880

[root@ATK-DLRV1126:/demo]# ./infer yolov8m_float16.tflite

Inference Time in ms: 51716

第 10 章和第 11 章，我们将分别讨论用 NPU 和 GPU 加速端侧模型推理。

第 10 章

NPU 推理加速
——无人机采茶机器人研发实践

无人机应用在茶园给我们带来很多便利，如借助无人机将山上采摘的新鲜茶叶及时运送到山下，通过无人机日常巡查茶园中茶叶生长状况，等等。

本章介绍的无人机采茶机器人，可高效实现茶叶的采摘，如图10-1所示。

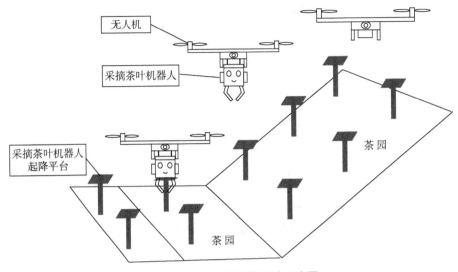

图 10-1　无人机采茶机器人示意图

乍一听，这是科幻小说吗？是笔者脑洞大开吗？不是！这是明天的精细化采茶模式（目前处于设想阶段），它无惧山高路远，能全天候采茶。

无人机采茶机器人包括无人机和采茶机器人两个主模块，它们的组合模式为登月模式。关于登月模式，具体解释如下：

无人机将采茶机器人载送到茶园的指定起降平台（这里的桩基指采摘茶叶机器人的降落台，数量为多个）后，然后控制采茶机器人着陆在起降平台上，后无人机和采茶机器人分离。无人机返回到服务站进入休息状态，其间可完成充电。

采茶机器人通过3D视觉模块实现茶树的芽叶的精准定位，然后控制采摘机械臂去采摘芽叶并回收。

采茶机器人完成起降平台范围的芽叶采摘后，给无人机发送信息指令。无人机收到指令后，飞来和采茶机器人对接，然后将其运送到下一个起降平台，继续重复采茶操作，或者运送到服务站，给采茶机器人充电和卸下茶叶。

它的主要创新包括以下几点：

（1）无人机和采茶机器人是一对多关系。

（2）登月模式可解决无人机的电池续航及无人机螺旋桨产生的大气流对茶叶的影响等问题。

（3）起降平台为茶园的基础配置，其成本很低，无人机采茶机器人批量化生产的成本也将很低。这种起降平台有很多的变体,也可以设计为沿着茶园中一行行的轨道模式。

(4)这种采茶模式高效(连续工作),具有高观赏性。

上文介绍的芽叶定位,我们采用的技术方案中涉及目标检测和NPU加速模型推理技术。

本章主要讨论如何使用NPU加速目标检测模型推理,内容涉及模型格式转换和验证、模型量化误差评估、NPU推理性能和内存使用情况评估等。本章中的NPU以RV1126芯片中内置的2T算力NPU核为测试平台,目标检测模型以与第9章中相同的模型为评估对象。

准备工作如下:

第一步,创建Python=3.8虚拟环境。

Python虚拟环境可以实现在一台机器上提供多个独立的隔离运行环境,避免不同的项目之间的Python版本解释器和依赖库的冲突。我们使用Conda可以很轻松地实现创建和管理Python虚拟环境,其创建命令为conda create -n env_name python=x.x,创建好的虚拟环境的文件env_name可以在Anaconda安装目录envs文件夹下找到。以下是具体的创建步骤。

```
$ conda create -n npu_rknn_py38 python=3.8
#
# To activate this environment, use
#
#     $ conda activate npu_rknn_py38
#
# To deactivate an active environment, use
#
#     $ conda deactivate
```

激活创建的虚拟环境:

$ source activate npu_rknn_py38

第二步,安装RKNN-Toolkit开发工具套件(在这里我们以版本v1.7.5为例)。

关于RKNN-Toolkit在官方文档《RKNN-Toolkit用户使用指南》和《RKNN Toolkit快速上手指南》(文档位置为rknn-toolkit-master-v1.7.5/doc)中有详细的描述。在这里参考以上文档简单介绍如下:

RKNN-Toolkit是为用户提供在PC、Rockchip NPU平台上进行模型转换、推理以及性能评估的开发工具套件。用户通过该工具套件中提供的Python接口可以便捷地实现基于Rockchip NPU平台的模型部署相关工作,主要包括以下功能:

① 将TensorFlow/TensorFlow Lite/ONNX/PyTorch/Keras等模型转为RKNN模型

（RK1126 NPU 必须使用 RKNN 模型进行推理）。

② 将浮点模型量化为定点模型，其中量化方法支持非对称量化（asymmetric_quantized-u8）、动态定点量化（dynamic_fixed_point-i8/dynamic_fixed_point-i16）等。

③ 将 RKNN 模型部署到指定的 NPU 核，并进行推理。

④ 评估模型在指定 NPU 核中的性能。

⑤ 评估模型运行时的系统内存使用情况。

⑥ 模型预编译。通过模型预编译生成的 RKNN 模型用于减少 NPU 加载模型的时间。预编译的 RKNN 模型只能在 NPU 硬件中运行。

⑦ 模型分段（用于多模型同时运行的场景，比如目标检测模型和语义分割模型需要同时运行在 NPU 中的场景）。将单个模型分成多段在 NPU 上执行，用于调节多个模型占用 NPU 的时间，避免因为一个模型占用太多的 NPU 时间，而使其他模型无法及时执行。

⑧ 自定义算子功能。当模型含有 RKNN-Toolkit 不支持的算子（operator），模型转换将失败。针对这种情况，RKNN-Toolkit 提供自定义算子功能，允许用户自己实现相应的算子，从而使模型能正常转换和运行。

⑨ 量化精度分析。保存浮点模型、量化模型推理中每一层的中间结果，并用欧式距离和余弦距离评估它们的相似度，从而评估模型量化的误差。

⑩ 可视化功能。简化用户操作步骤。

⑪ 模型优化等级调整。RKNN-Toolkit 在模型转换过程中会对模型进行优化，默认的优化选项可能会对模型精度或性能产生一些影响。通过设置优化等级，可以关闭部分或全部优化选项。

⑫ 模型加密功能。使用指定的加密等级实现 RKNN 模型加密。RKNN 模型的加密是在 NPU 驱动中完成的，所以加载加密模型的方式和加载普通 RKNN 模型方式一样，NPU 驱动会自动对其进行解密。

⑬ 可导出 RKNN 模型中每一层的量化参数。

我们通过网址 https://github.com/rockchip-linux/rknn-toolkit/releases 下载 rknn-toolkit-v1.7.5-packages 软件包，然后进入软件包的子路径 rknn-toolkit-v1.7.5-packages/packages，再通过命令 pip install rknn_toolkit-1.7.5-cp38-cp38-linux_x86_64.whl -i https://pypi.tuna.tsinghua.edu.cn/simple/ 安装 RKNN-Toolkit，相关准备工作完成。

10.1 基于 ONNX 格式模型转换和验证

在这里的模型转换指将 ONNX 模型转换成 Rockchip NPU 推理的 RKNN 模型，具体的步骤如下（以 yolov8s.pt 模型为例）。

第一步，克隆代码仓库。

git clone https://github.com/airockchip/ultralytics_yolov8.git

ultralytics_yolov8$ git log

commit 5b7ddd8f821c8f6edb389aa30cfbc88bd903867b（HEAD -> main，origin/main，origin/HEAD）

第二步，onnx 格式模型转换。

先将代码工程 ./ultralytics/cfg/default.yaml 中的配置第 8 行修改如下：

model: yolov8s.pt

然后运行命令 pip install -r requirements.txt -i https://pypi.tuna.tsinghua.edu.cn/simple/ 安装相关的依赖包后，运行以下命令实现 yolov8s.pt 转化为 yolov8s.onnx 格式模型。

ultralytics_yolov8$ python ./ultralytics/engine/exporter.py

转化日志如下：

Ultralytics YOLOv8.0.151 🚀 Python-3.8.19 torch-2.2.2+cu121 CPU（）

YOLOv8s summary（fused）：168 layers, 11156544 parameters, 0 gradients

PyTorch: starting from 'yolov8s.pt' with input shape（16，3，640，640）BCHW and output shape（s）((16，64，80，80)，(16，80，80，80)，(16，1，80，80)，(16，64，40，40)，(16，80，40，40)，(16，1，40，40)，(16，64，20，20)，(16，80，20，20)，(16，1，20，20))（21.5 MB）

RKNN: starting export with torch 2.2.2+cu121...

RKNN: feed yolov8s.onnx to RKNN-Toolkit or RKNN-Toolkit2 to generate RKNN model.

Refer https://github.com/airockchip/rknn_model_zoo/tree/main/models/CV/object_detection/yolo

RKNN: export success ✓ 0.9s, saved as 'yolov8s.onnx'（42.6 MB）

Export complete（7.2s）

Results saved to /······/ultralytics_yolov8

Predict: yolo predict task=detect model=yolov8s.onnx imgsz=640

Validate: yolo val task=detect model=yolov8s.onnx imgsz=640 data=coco.yaml

Visualize: https://netron.app

yolov8s.onnx 模型的网络结构如图 10-2 所示。

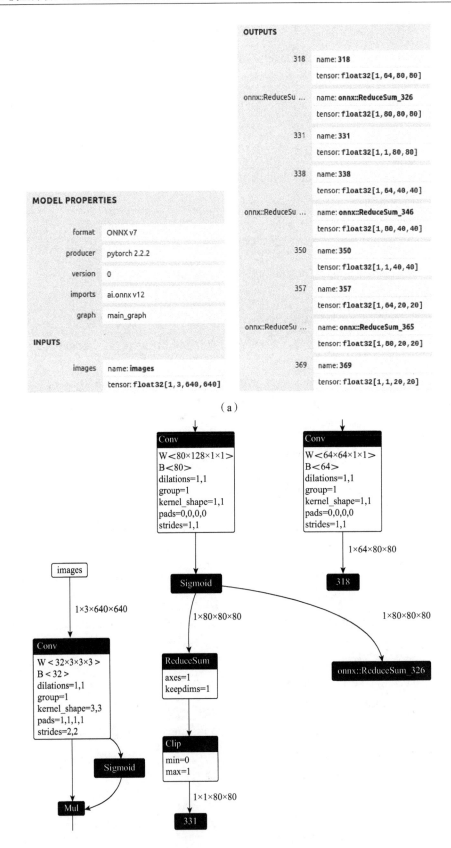

(a)

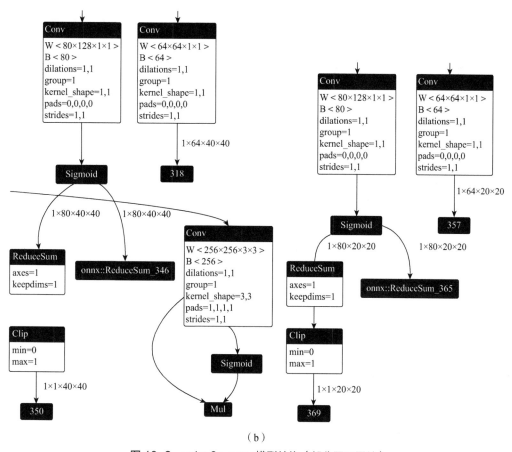

（b）

图 10-2 yolov8s.onnx 模型结构（部分展示图片）

第三步，.rknn 格式模型转换。

首先，激活 npu_rknn_py38 虚拟环境和克隆 rknn_model_zoo 代码仓库。

git clone https://github.com/airockchip/rknn_model_zoo.git

其中在 rknn_model_zoo/examples/yolov8/python 路径中 convert.py 文件用于实现 yolov8 模型的 .rknn 格式转换，它的关键代码如下：

```
if __name__ == '__main__':
    model_path = "yolov8s.onnx"
    platform = ['rv1126']
    do_quant = True
    output_path = "yolov8s.rknn"
    # Create RKNN object
    rknn = RKNN（verbose=True）

    # Pre-process config
```

```python
    print('--> Config model')
    rknn.config(mean_values=[[0, 0, 0]],
                std_values=[[255, 255, 255]],
                target_platform=platform)
    print('done')

    # Load model
    print('--> Loading model')
    ret = rknn.load_onnx(model=model_path)
    if ret != 0:
        print('Load model failed!')
        exit(ret)
    print('done')

    # Build model
    print('--> Building model')
    ret = rknn.build(do_quantization=do_quant,
                     dataset=DATASET_PATH)
    if ret != 0:
        print('Build model failed!')
        exit(ret)
    print('done')

    # Export rknn model
    print('--> Export rknn model')
    ret = rknn.export_rknn(output_path)
    if ret != 0:
        print('Export rknn model failed!')
        exit(ret)
    print('done')

    # Release
    rknn.release()
```

运行命令 python convert.py ../model/yolov8s.onnx rv1126 实现 yolov8s.onnx 到 yolov8s.rknn 格式模型的转换。转换的 yolov8s.rknn 模型结构如图 10-3 所示。

第 10 章 NPU 推理加速——无人机采茶机器人研发实践

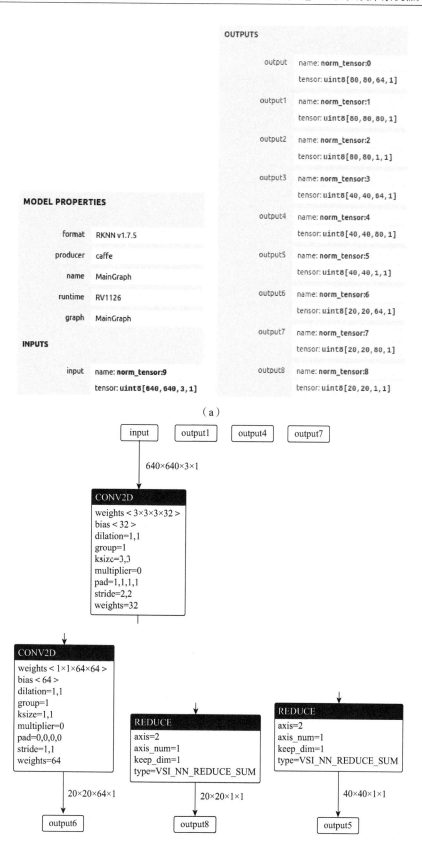

（a）

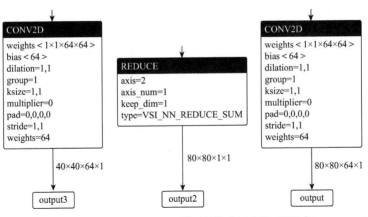

图 10-3 yolov8s.rknn 模型结构（部分展示图片）

第四步，.rknn 格式模型验证。

我们在代码文件的 /rknn_model_zoo/examples/yolov8/python 路径下运行命令 python yolov8.py --model_path'yolov8s.rknn'--target'rv1126'--img_show，得到 yolov8s.rknn 模型在 RV1126 平台中的推理效果，如图 10-4 所示。

图 10-4 yolov8s.rknn @RV1126 平台推理[①]

10.2 模型量化误差评估

首先在代码文件 /rknn_model_zoo/examples/yolov8/python 路径下运行命令 python yolov8.py --model_path yolov8s.onnx --coco_map_test --img_save，评估 yolov8s.onnx 模型基于 COCO 数据集的推理效果，并作为评估基准，得到的推理结果如下：

---

① 照片拍摄于前往西递、宏村的路上，路上的山坡茶园给笔者带来了很多的启发。无人机采茶机器人的想法也是出于此。

```
Average Precision (AP) @ [ IoU=0.50:0.95 | area=   all   | maxDets=100 ] = 0.391
Average Precision (AP) @ [ IoU=0.50      | area=   all   | maxDets=100 ] = 0.524
Average Precision (AP) @ [ IoU=0.75      | area=   all   | maxDets=100 ] = 0.426
Average Precision (AP) @ [ IoU=0.50:0.95 | area= small   | maxDets=100 ] = 0.193
Average Precision (AP) @ [ IoU=0.50:0.95 | area=medium   | maxDets=100 ] = 0.435
Average Precision (AP) @ [ IoU=0.50:0.95 | area= large   | maxDets=100 ] = 0.566
Average Recall    (AR) @ [ IoU=0.50:0.95 | area=   all   | maxDets=  1 ] = 0.314
Average Recall    (AR) @ [ IoU=0.50:0.95 | area=   all   | maxDets= 10 ] = 0.444
Average Recall    (AR) @ [ IoU=0.50:0.95 | area=   all   | maxDets=100 ] = 0.450
Average Recall    (AR) @ [ IoU=0.50:0.95 | area= small   | maxDets=100 ] = 0.218
Average Recall    (AR) @ [ IoU=0.50:0.95 | area=medium   | maxDets=100 ] = 0.494
Average Recall    (AR) @ [ IoU=0.50:0.95 | area= large   | maxDets=100 ] = 0.647
map  --> 0.39102777321587107
map50--> 0.5243398254349063
map75--> 0.4258175914261063
map85--> 0.4944908595354903
map95--> 0.6468566661526711
```

然后运行命令 python yolov8.py --model_path yolov8s.rknn --target 'rv1126' --coco_map_test --img_save 评估 yolov8s.rknn 模型的推理效果，得到的推理结果如下（图 10-5 展示部分推理效果图）：

```
Average Precision (AP) @ [ IoU=0.50:0.95 | area=   all   | maxDets=100 ] = 0.381
Average Precision (AP) @ [ IoU=0.50      | area=   all   | maxDets=100 ] = 0.513
Average Precision (AP) @ [ IoU=0.75      | area=   all   | maxDets=100 ] = 0.416
Average Precision (AP) @ [ IoU=0.50:0.95 | area= small   | maxDets=100 ] = 0.186
Average Precision (AP) @ [ IoU=0.50:0.95 | area=medium   | maxDets=100 ] = 0.423
Average Precision (AP) @ [ IoU=0.50:0.95 | area= large   | maxDets=100 ] = 0.554
Average Recall    (AR) @ [ IoU=0.50:0.95 | area=   all   | maxDets=  1 ] = 0.307
Average Recall    (AR) @ [ IoU=0.50:0.95 | area=   all   | maxDets= 10 ] = 0.434
Average Recall    (AR) @ [ IoU=0.50:0.95 | area=   all   | maxDets=100 ] = 0.439
Average Recall    (AR) @ [ IoU=0.50:0.95 | area= small   | maxDets=100 ] = 0.209
Average Recall    (AR) @ [ IoU=0.50:0.95 | area=medium   | maxDets=100 ] = 0.482
Average Recall    (AR) @ [ IoU=0.50:0.95 | area= large   | maxDets=100 ] = 0.634
map  --> 0.3813096906769016
map50--> 0.513038673933291
map75--> 0.4159382621935495
map85--> 0.482311621610949
map95--> 0.6339543968389261
```

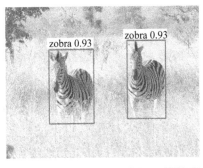
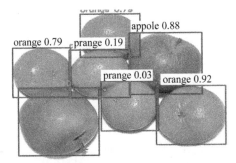
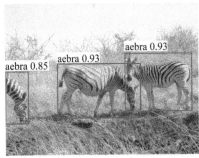

图 10-5　部分推理效果图 [①]

评估结论：

通过以上结果的对比分析，yolov8s.onnx 模型经过量化后得到的 yolov8s.rknn 模型的精度降低 0.1%。如何让精度损失更低？量化感知训练是一个很好的解决方案。基于 RKNN-Toolkit 工具套件如何实现量化感知训练？鉴于本书篇幅的限制，读者朋友可以参考《RKNN-Toolkit 用户使用指南》做进一步的测试。

10.3　NPU 推理性能和内存使用情况评估

（1）NPU 推理性能评估

RKNN-Toolkit 为我们提供了 .rknn 模型的推理性能评估接口 eval_perf，通过这个接口可以收集统计模型的运行耗时，包括每层耗时、总耗时、每帧的平均耗时。我们通过以下代码实现 yolov8s.rknn 模型在 RV1126 中 NPU 核推理性能评估。

```
ret=rknn.init_runtime（target='rv1126', device_id='******', perf_debug=True）
rknn.eval_perf（loop_cnt=10）# loop_cnt 为循环次数
```

具体的评估结果如图 10-6 所示。

---

① 图来源：COCO 开源数据集

```
Layer ID  Name                                              Operator      Uid   Time(us)
90        Conv_/model.0/conv/Conv_226_2                     CONVOLUTION   226   22482
0         Sigmoid_/model.0/act/Sigmoid_225_Mul_/mo          ACTIVATION    224   19748
          del.0/act/Mul_224_3
91        Conv_/model.1/conv/Conv_222_2                     CONVOLUTION   222   9947
1         Sigmoid_/model.1/act/Sigmoid_223_Mul_/mo          ACTIVATION    221   7727
          del.1/act/Mul_221_3
92        Conv_/model.2/cv1/conv/Conv_220_2                 CONVOLUTION   220   7125
2         Sigmoid_/model.2/cv1/act/Sigmoid_219_Mul          ACTIVATION    217   7485
          _/model.2/cv1/act/Mul_217_3
3         TensorCopy_3                                      SPECIAL_OP          600
4         TensorCopy_3                                      SPECIAL_OP          609
93        Conv_/model.2/m.0/cv1/conv/Conv_213_2             CONVOLUTION   213   1858
5         Sigmoid_/model.2/m.0/cv1/act/Sigmoid_210          ACTIVATION    209   2323
          _Mul_/model.2/m.0/cv1/act/Mul_209_3
94        Conv_/model.2/m.0/cv2/conv/Conv_203_2             CONVOLUTION   203   1810
6         Sigmoid_/model.2/m.0/cv2/act/Sigmoid_204          ACTIVATION    198   1602
          _Mul_/model.2/m.0/cv2/act/Mul_198_3
7         Add_/model.2/m.0/Add_196_1                        TENSOR_ADD    196   2307
95        Conv_/model.2/cv2/conv/Conv_188_2                 CONVOLUTION   188   15029
          ............

149       Conv_/model.22/cv2.2/cv2.2.1/conv/Conv_2          CONVOLUTION   29    814
          9_2
86        Sigmoid_/model.22/cv2.2/cv2.2.1/act/Sigm          ACTIVATION    20    465
          oid_30_Mul_/model.22/cv2.2/cv2.2.1/act/M
          ul_20_3
151       Conv_/model.22/cv2.2/cv2.2.2/Conv_11_2            CONVOLUTION   11    452
148       Conv_/model.22/cv3.2/cv3.2.0/conv/Conv_6          CONVOLUTION   63    777
          3_2
85        Sigmoid_/model.22/cv3.2/cv3.2.0/act/Sigm          ACTIVATION    51    885
          oid_52_Mul_/model.22/cv3.2/cv3.2.0/act/M
          ul_51_3
150       Conv_/model.22/cv3.2/cv3.2.1/conv/Conv_3          CONVOLUTION   39    933
          9_2
87        Sigmoid_/model.22/cv3.2/cv3.2.1/act/Sigm          ACTIVATION    28    352
          oid_40_Mul_/model.22/cv3.2/cv3.2.1/act/M
          ul_28_3
152       Conv_/model.22/cv3.2/cv3.2.2/Conv_19_2            CONVOLUTION   19    147
88        Sigmoid_/model.22/Sigmoid_2_10_3                  ACTIVATION    10    759
89        TensorReduceSum_1                                 SPECIAL_OP          1097
Total Time(us): 553038
FPS: 1.81
===================================================================
=================== Operator time-consuming statistics ============
OP            Call Number     Total Time(us)    Time Ratio(%)
CONVOLUTION   63              275602            49.83
ACTIVATION    60              250962            45.38
SPECIAL_OP    21              14879             2.69
TENSOR_ADD    6               7991              1.44
POOLING       3               3604              0.65
===================================================================
```

图 10-6　yolov8s.rknn @RV1126 NPU 推理性能评估

以上评估中的 perf_debug 参数设置为 True，是为了在评估中可以收集每一层的耗时信息，所以 NPU 驱动会在每一层推理前后插入一些代码，从而导致总耗时比实际耗时长。

我们将 perf_debug 参数设置为 False，得到的评估信息如下：

perf_debug=False

Average inference Time（us）: 74307.8

FPS: 13.46

可以发现，FPS 明显变大。

（2）内存使用情况评估

我们通过以下方法实现 RKNN 模型运行时的内存使用情况的评估。

ret=rknn.init_runtime（target='rv1126', device_id='******', eval_mem=True）
rknn.eval_memory（）

评估结果如下，其中 maximum allocation 为内存使用峰值，表示从模型运行开始到结束过程中内存的最大分配值；total allocation 为总内存值，表示模型运行期间分配的所有内存之和。

```
======================================================
              Memory Profile Info Dump
======================================================
```

System memory（非 NPU 驱动分配的系统内存，包括为模型、输入数据等在系统中分配的内存）：

    maximum allocation : 39.18 MiB

    total allocation　 : 473.52 MiB

NPU memory（NPU 驱动在运行期间使用的内存）：

    maximum allocation : 45.59 MiB

    total allocation　 : 45.60 MiB

Total memory（System memory 和 NPU memory 内存之和）：

    maximum allocation : 84.77 MiB

    total allocation　 : 519.12 MiB

INFO: When evaluating memory usage, we need consider the size of model, current model size is: 10.81 MiB

```
======================================================
```

通过以上的讨论，我们已经初步了解了如何使用 NPU 去加速推理一个模型，其中涉及模型的转换、量化误差评估、推理性能和内存使用情况评估。但是这只是初步了解，读者朋友还需要在实际项目中实践，遇到问题解决问题，让 NPU 加速推理模型的方案在实际产品开发中发挥价值。当然其中的乐趣，只有亲历才能体会，也是非常奇妙。

第 11 章

端侧 GPU 硬件加速模型推理
——智能水下摄像机器人研发实践

智能水下摄像机器人（Underwater Camera Robot）实现在水下拍照和录制视频（配有智能补光设备）、智能感知和机械臂操作等功能，可应用在水下智能监控、水下搜救、水下勘察、安全巡检和科研探索等领域。

举例说明：通过智能水下摄像机器人来监控鱼群生长状况；用于检查水库大坝、桥梁（水下桥墩）和管道的状况，排除隐患；用于检查船舶底部的附着物、推进器、水下锚等情况。

目前，水下摄像会遇到很多挑战，如水域深浅、水流大小、水质浑浊（水下水质浑浊度度量算法）、水对光的吸收、光的散射和衰减、水下光线昏暗（水下光线度量）等等，还会遇到设备自身在水下的腐蚀以及受到水压、温度、撞击等挑战。所以我们的设计方案中会考虑到：采用智能雨刷去避免镜头遮挡和模糊，异常声音检测，异常温度检测和过热保护，设备撞击检测，设备进水和水汽检测，以及本章中讨论的水下图像增强（Underwater Image Enhancement）算法。

相机在水下和水上拍的物体会不一样，如水下图像对比度低、色彩失真和颜色偏移等，特别当水质浑浊时，成像更加困难。如何让相机在水下拍得更清晰，这是图像增强算法需要做的，如亮度和色彩矫正、去噪等。

本章以图像增强模型和水下目标检测模型（YOLOv8）为例，讨论如何通过端侧GPU加速模型推理，内容包括模型转换、模型部署和评估等。主要以高通骁龙 7 Gen1（4×2.36GHz Cortex-A710，4×1.8GHz Cortex-A510，ISA ARMv8.2-A，集成 Adreno 644 GPU）为测试平台。

11.1 水下图像增强模型——Transformer 模型端侧部署评估

Transformer 模型在很多任务中都能够取得非常好的效果，如 Sora 底层的秘密少不了它，很多大模型都是基于 Transformer 搭建的。本小节讨论的水下图像增强模型也是以 Transformer 为基础模型，实现水下图像对比度的提升和色彩的还原（图 11-1），如数据集 EUVP（Enhancing Underwater Visual Perception，增强水下视觉感知）中 Ground Truth（GT）。

图 11-1　EUVP 数据集[①]

11.1.1　水下图像增强模型网络结构和可视化分析

水下图像增强模型的网络结构如图 11-2 所示。图（a）为水下图像增强模型，将图像归一化后，首先输入到模型的头部（卷积层模块），用于提取浅层特征，然后输入到模型中部，用于提取更高层次的特征，最后输入到模型的尾部（卷积层模块），用于构建恢复图像。

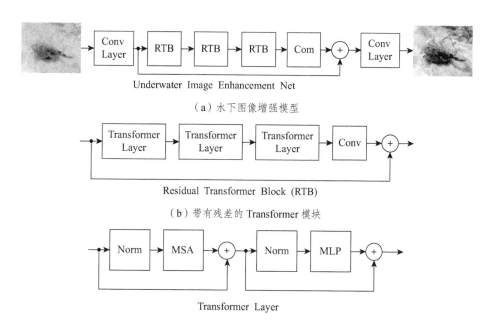

（a）水下图像增强模型

（b）带有残差的 Transformer 模块

（c）Transformer 基本模块 [MSA（Multi-Head Self-Altention）为多头自注意力机制]

图 11-2　水下图像增强模型的网络结构

MSA 的网络结构如图 11-3（a）所示，它是图像增强模型中计算量最大的部分，如果部署到端侧，就需要做性能、精度和功耗的优化。它的轻量化解决方案是下面将要讨论的内容。

attentionless Transformers 是苏黎世联邦理工学院（ETH Zurich）针对 Transformer 模

---

① 数据地址：https://irvlab.cs.umn.edu/resources/euvp-dataset

块提出的 MLP 替换方案。MLP（Multilayer Perceptron，多层感知器）是一种前馈神经网络（Feed-Forward Neural Network，FFN），一般只使用全连接层和激活函数作为构建算子。它的输入和输出必须是一维向量（张量），如果输入的不是一维向量，需要先通过 Flatten 层进行展平处理。具体提到的 MLP 替换方案有四种不同程度的替换。

方案 1：用 MLP 替换多头注意力模块（MHA），如图 11-3（b）所示。
方案 2：用 MLP 替换 MHA 模块和残差连接，如图 11-3（c）所示。
方案 3：用单独的 MLP 替换 MHA 模块中每个头，如图 11-3（d）所示。
方案 4：MLP 替换编码器层，如图 11-3（e）所示。

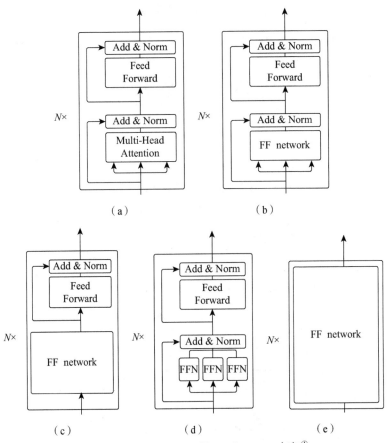

图 11-3 attentionless Transformers 方案[1]

CloFormer 是一种轻量化的 Transformer 方案，关键模块为 Clo Block 和 ConvFFN。
Clo Block 由一个全局分支（Global）和一个局部分支（Local）组成，如图 11-4（a）所示。Clo Block 的输入经过 LN 归一化后分两路，分别输入到全局和局部分支。两个分支

---

[1] 图来源：Vukasin Bozic, Danilo Dordevic, Daniele Coppola, et al. Rethinking Attention: Exploring Shallow Feed-Forward Neural Networks as an Alternative to Attention Layers in Transformers. arXiv preprint arXiv: 2311.10642，2023.

中,都先通过各自的 FC(全连接层)做线性变换,然后分三路作为 Q、K 和 V 量。在全局分支中,K 和 V 量通过对应的 pool 下采样减少计算量,然后对 Q、K 和 V 进行标准的自注意力操作,提取低频全局信息(低频分量为图像亮度和颜色变化平缓区域)。在局部分支中,引入 AttnConv 卷积操作,利用 DWconv 权重共享和上下文感知权重进行局部感知,提取高频信息(高频分量为图像细节:边缘和纹理)。关键过程为将 Q、K 和 V 量分别输入到对应的 DWconv 中提取局部信息,然后将 Q 和 K 点乘等一系列处理进行上下文感知。

ConvFFN 在 GELU 激活函数后使用 DWconv 操作提取局部信息,如图 11-4(b)所示。

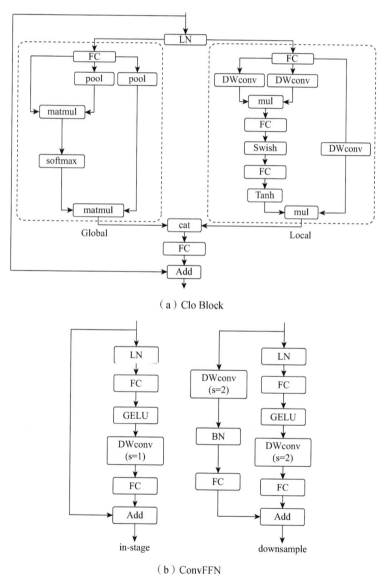

图 11-4 CloFormer 模型中关键模块[①]

———————
① 图来源:Rethinking Local Perception in Lightweight Vision Transformer

11.1.2 基于 QNN 推理框架 GPU 核部署评估

将上文讨论的模型部署在端侧 GPU 核，在本次评估中使用高通骁龙 7 Gen1 平台。我们使用 QNN 推理框架来部署。首先完成 QNN 的准备工作，关键步骤如下。

安装虚拟环境：

$ conda create -n qnn python=3.8

QNN 配置：

$ QNN_SDK_ROOT=/opt/qcom/aistack/qairt/2.21.0.240401

$ {QNN_SDK_ROOT}/bin/check-python-dependency

$ sudo bash ${QNN_SDK_ROOT}/bin/check-linux-dependency.sh

$ source ${QNN_SDK_ROOT}/bin/envsetup.sh

［INFO］AISW SDK environment set

［INFO］QNN_SDK_ROOT: /opt/qcom/aistack/qairt/2.21.0.240401

［INFO］SNPE_ROOT: /opt/qcom/aistack/qairt/2.21.0.240401

安装相关依赖包：

pip install torch==1.13.1 torchvision==0.14.1 -i https://pypi.tuna.tsinghua.edu.cn/simple/

pip install tensorflow==2.10.1 -i https://pypi.tuna.tsinghua.edu.cn/simple/

pip install onnxruntime==1.11.1 onnx==1.11.0 onnxsim==0.4.5 -i https://pypi.tuna.tsinghua.edu.cn/simple/

　　模型格式转换，得到的 former.pb 模型的输入输出节点名称如下：

［<tf.Tensor＇x:0＇ shape=（None, 640, 640, 3）dtype=float32>］

frozen_func.outputs:

［<tf.Tensor＇Identity:0＇ shape=（None, 320, 320, 3）dtype=float32>］

$ qnn-tensorflow-converter -i former.pb -d x 1, 640, 640, 3 --out_node Identity --output_path former.cpp

端侧模型运行测试，在路径 /opt/qcom/aistack/qairt/2.21.0.240401/examples/QNN/NetRun/android 下分别运行 bash android-qnn-net-run.sh -b cpu 和 bash android-qnn-net-run.sh -b gpu，得到的评估结果如表 11-1 所示。

表 11-1　端侧模型运行测试结果

硬件核	模型推理结果
CPU	$ qnn-profile-viewer --input_log qnn-profiling-data_0.log Time Scale: 1e-06 Epoch Timestamp: 1714478144016777 Steady Clock Timestamp: 326848269130

续表

硬件核	模型推理结果
CPU	Generated using: qnn-profile-viewer v2.21.0.240401120655_85612 qnn-net-run v2.21.0.240401120655_85612 Backend v2.21.0.240401120655_85612 Qnn Init/Prepare/Finalize/De-Init/Execute/Lib-Load Statistics: -- Init Stats: ----------- 　NetRun: 11,050 us Compose Graphs Stats: -------------- 　NetRun: 9,397 us Finalize Stats: --------------- Graph 0（former）： 　NetRun: 1,649 us 　　Backend（GRAPH_FINALIZE）：1,641 us De-Init Stats: -------------- 　NetRun: 140,324 us 　Backend（null）：1 us Execute Stats（Average）： ------------------------ Total Inference Time: --------------------- Graph 0（former）： 　NetRun: 3,494,109 us 　　Backend（GRAPH_EXECUTE）：3,490,504 us Execute Stats（Overall）： ------------------------ 　NetRun IPS（includes IO and misc. time）：0.2857 inf/sec

续表

硬件核	模型推理结果

GPU	$ qnn-profile-viewer --input_log qnn-profiling-data_0.log
	Time Scale: 1e-06
	Epoch Timestamp: 1714477871879944 Steady Clock Timestamp: 326576132297
	Generated using:
	qnn-profile-viewer v2.21.0.240401120655_85612
	qnn-net-run v2.21.0.240401120655_85612
	Backend v2.21.0.240401120655_85612
	
	Qnn Init/Prepare/Finalize/De-Init/Execute/Lib-Load Statistics:
	-- 	
	Init Stats:
	----------- 	
	  NetRun: 3,076,609 us
	
	Compose Graphs Stats:
	--------------------- 	
	  NetRun: 28,585 us
	
	Finalize Stats:
	--------------- 	
	Graph 0（former）:
	  NetRun: 3,048,009 us
	  Backend（QnnGraph_finalize）: 3,047,995 us
	
	De-Init Stats:
	-------------- 	
	  NetRun: 125,405 us
	  Backend（QnnContext_free）: 125,398 us
	
	Execute Stats（Average）:
	------------------------ 	
	Total Inference Time:
	--------------------- 	
	Graph 0（former）:
	  NetRun: 803,305 us
	  Backend（QnnGraph_execute）: 803,284 us
	
	Execute Stats（Overall）:
	------------------------ 	
	  NetRun IPS（includes IO and misc. time）: 1.2341 inf/sec

11.2 水下目标检测模型——YOLOv8 模型端侧部署评估

水下目标检测让水下摄像机器人具有感知的能力，如图 11-5 所示，我们通过水下摄像机器人可以很方便地实现水下的环境观察。

图 11-5 水下目标检测[①]

基于端侧设备的 GPU 核实现 YOLOv8 模型的加速推理是本小节讨论的内容。首先，使用命令 qnn-onnx-converter --input_network yolov8s.onnx --output_path yolov8s.cpp 将 yolov8s.onnx 模型转化为 QNN 部署的模型格式。我们分别在 CPU 和 GPU 核评估了 yolov8s.bin 模型的推理性能。

（1）CPU 核

（qnn）user@user:/opt/qcom/aistack/qairt/2.21.0.240401/examples/QNN/NetRun/android/output-cpu$ qnn-profile-viewer --input_log qnn-profiling-data_0.log

Time Scale: 1e-06

Epoch Timestamp: 1714138611261918 Steady Clock Timestamp: 207610373565

Generated using:

qnn-profile-viewer v2.21.0.240401120655_85612

qnn-net-run   v2.21.0.240401120655_85612

Backend   v2.21.0.240401120655_85612

Qnn Init/Prepare/Finalize/De-Init/Execute/Lib-Load Statistics:

Init Stats:

---

① 图来源：https://conservancy.umn.edu/handle/11299/214865

NetRun: 86101 us

Compose Graphs Stats:

NetRun: 59689 us

Finalize Stats:

Graph 0（yolov8s）：

NetRun: 26407 us

Backend（GRAPH_FINALIZE）：26400 us

De-Init Stats:

NetRun: 28461 us

Backend（null）：0 us

Execute Stats（Average）：

Total Inference Time:

Graph 0（yolov8s）：

NetRun: 834861 us

Backend（GRAPH_EXECUTE）：830842 us

Execute Stats（Overall）：

NetRun IPS（includes IO and misc. time）：1.1900 inf/sec

（2）GPU 核

（qnn）user@user:/opt/qcom/aistack/qairt/2.21.0.240401/examples/QNN/NetRun/android/output-gpu$ qnn-profile-viewer --input_log qnn-profiling-data_0.log

Time Scale: 1e-06

Epoch Timestamp: 1714138736394245 Steady Clock Timestamp: 207735505892

Generated using:

qnn-profile-viewer v2.21.0.240401120655_85612

qnn-net-run   v2.21.0.240401120655_85612

Backend   v2.21.0.240401120655_85612

Qnn Init/Prepare/Finalize/De-Init/Execute/Lib-Load Statistics:

Init Stats:

 NetRun: 2725194 us

Compose Graphs Stats:

 NetRun: 63469 us

Finalize Stats:

Graph 0（yolov8s）：

 NetRun: 2661716 us

 Backend（QnnGraph_finalize）: 2661707 us

De-Init Stats:

 NetRun: 29062 us

 Backend（QnnContext_free）: 29052 us

Execute Stats（Average）：

Total Inference Time:

Graph 0（yolov8s）：

 NetRun: 297398 us

 Backend（QnnGraph_execute）: 297377 us

Execute Stats（Overall）：

 NetRun IPS（includes IO and misc. time）: 3.2812 inf/sec

第 12 章

安全智能——以隐私 OCR 为实例

大模型已经很火，还要继续火，瞧：这边忙着落地，那边忙着发 paper。目前阶段的大模型的模型参数量是以亿为单位（用字母 B（Billion）表示，1B 等于 10 亿），从亿级到百万亿级，其训练需要的数据量也是惊人的，往往从几百个 G 到几百个 T，甚至更多。当然得到的收益也很惊人，大模型所体现出来的智能也让人叹为观止。不过，大模型的能力还需要去优化。比如大语言模型有时候也会发现是"睁眼说瞎话"，但是我们相信这些大模型都是在不断进化中。说到模型的进化，就离不开数据或语料，您希望您的聊天记录、生活照片被用作模型的训练数据吗？当然不希望和不允许，这就涉及一个非常重要的问题：AI 服务的安全和智能。

我们总是说：用户更在乎您的产品能不能帮他们解决问题，而不关心您的技术。没错，但是忽视 AI 服务安全（这里主要讨论隐私安全），往往会给用户带来隐私泄露的问题，这能叫解决问题吗？

在本章中以一款产品——隐私 OCR，为实例去讨论我们该如何去开发更好的 AI 产品。安全和智能是左手和右手的关系，缺一不可。

OCR（Optical Character Recognition，光学字符识别）是指通过计算机视觉算法实现影像文件（比如证件、票据、文本、车牌、卡片、海报等）中字符（印刷体和手写体）的定位、识别和结构化的过程。所谓结构化，是指将识别出来的结果建立一定的键值对应关系，比如{"姓名"："猪八戒"}、{"地址"："高老庄"}等。通过 OCR 字符识别技术进行自动识别，可以快捷高效地实现自动信息录入和分类，减少用户手动输入的频次，极大提升用户体验。

笔者认为：广义的 OCR 是指以 OCR 相关联的技术为依托的一系列的场景应用，比如全文识别、自定义模板识别、智能模板匹配、证件分类、手写字体识别、印章识别、证件伪造检测、拍照增强等等。

12.1　OCR 的四个矛盾

偶尔听到有人说 OCR 简单！！！真的简单吗？

笔者认为技术往往可以说简单（这里的简单是指一般的实现，或者是理想的实现），但是基于不同的业务场景的落地却是非常难的。OCR 从产品定义、开发、打磨、部署到运营，笔者认为目前主要存在以下几个矛盾（鱼和熊掌的问题）：

（1）算法复杂度和端侧资源的矛盾

算法复杂度包括时间复杂度和空间复杂度，前者受到端侧算力的制约，后者会受到端侧内存的限制。算法复杂度直接影响到用户的体验感，比如识别一张证件中的字符信息，如果用户需要等待超过 5 秒钟，那么用户肯定会抓狂。

以文档扫描为例，其中自动纠偏和寻边裁剪的功能（图 12-1），这个算法经过优化后部署在手机端，用户不需要联网也可以很方便地使用这个功能，这个也叫离线服务。

图 12-1　智能寻边裁剪

当我们用某一款 APP 去识别一个表格的时候，如果关闭手机的网络，可能会发现这个功能不能使用。因为这个算法是部署在云端（或者端云协作）。为什么不部署在端侧呢？其中一个原因就是端侧资源的限制。

（2）产品功能和算法矛盾

产品功能定义的多和复杂，必然导致增加算法的复杂度。

对于图片中复杂的版面分析，如图 12-2 所示，其中包含字符、表格、图片等元素，往往也对算法提出不小的挑战。

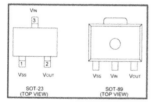
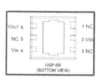
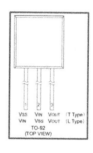

图 12-2　元素较多的图片[①]

字符定位模块中要求支持如图 12-3 中右边所示的不同的角度字符定位，这必然会增加字符定位算法的复杂度。

---

① 图来源：XC6206 芯片手册

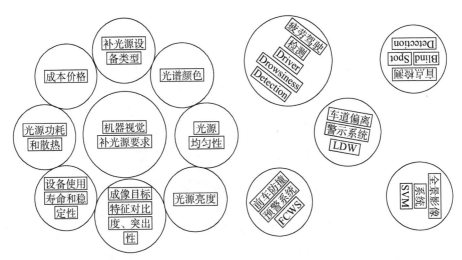

图 12-3　OCR 字符定位（左边为水平方向字符定位，右边为倾斜方向字符定位）

字符识别模块要求支持手写体（图 12-4）、生僻字、异体字、艺术字、模糊字、AIGC 光影字等识别，这对识别模型是不小的挑战。

识别结果
土鸡蛋16.8元/斤

图 12-4　手写体识别（休息一会，写代码后，去煎一个鸡蛋犒劳一下自己）

（3）产品理解力和产品的矛盾

笔者一直认为产品理解力和数据、算法、算力一起构成 AI 的四要素，这不同于当前的三要素的说法。产品理解力决定产品的上限。如模型的时延很大，会影响用户的体验感，我们通过改变业务的流程将原先串行的流程优化为并行的，让用户感觉不到模型的等待，这样如果同样可以满足业务的需求，是不是很好？

（4）隐私数据和云端服务不透明的矛盾

安全是首要需求，是必须严格遵守的法则。这里的安全包括数据安全和服务安全两个方面，它们具体涉及服务设备、终端、逻辑、通信、非法攻击、隐私等。比如金融开户、电商交易等场景中，安全高于一切。数据和服务的安全是体验感的重要部分。公安部参与起草的《信息安全技术——远程人脸识别系统技术要求》和工业和信息化部发布的《App 收集使用个人信息最小必要评估规范》(提出以"最小必要化"原则去采集用户证件、人脸等个人信息)足以说明安全的重要性。用户对于自己的数据,应该保障其知悉权,用户有权利对自己的数据可追溯。数据的滥用、数据垄断、数据的归属、数据的授权边界、数据的信息安全等等,都是我们需要思考和解决的问题。

模型端侧部署当然是目前最佳的保护用户隐私的方案,但是鉴于以下几个方面的原因,会选择端云协作或者云端部署的方案。

a. 端侧算力等硬件资源的约束,这点在矛盾(1)中已经讨论。

b. 商业模式。

对于模型部署,在云端和端侧涉及收费的模式。部署在云端,可以采取按照请求次数收费;如果部署在端侧,可能就需要用户一次支付相对比较高的费用,对于用户来说,会不会买单是一个问题。

模型的开发,一般都是基于服务端开发,开始关注模型的精度,而不再考虑端侧的性能。为了服务的快速推广,云端的部署方案往往会先走一步。

如果采取模型的端侧部署,会需要更多的研发投入,比如模型压缩、模型加密、不同平台的适配等。

c. 模型安全。

模型安全是保护供应商的知识产权以及防止模型的窃取。在云端部署可以很好地保护自己的模型,而在端侧部署,就必须要考虑模型的加固,这个对技术往往也是一笔很高的投入。

鉴于以上的讨论,我们的产品隐私 OCR 该如何设计呢?

先讨论有关 OCR 的主要技术方法。

图 12-5 所示为传统 OCR 的主要流程,其中图(a)为整体流程,图(b)是经过图像大小归 化、RGB 转灰度图、高斯滤波、图像增强、直方图均衡化和二值化处理处理后得到的结果;图(c)为经过 Canny 边缘检测、寻找边缘轮廓、筛选最大轮廓和角点定位等一系列的处理后得到的识别图像中主体区域的定位。

(a)

（b） （c）

图 12-5 传统 OCR 技术

传统 OCR 技术容易受到如图 12-6 所示的图像模糊、光斑遮挡、低亮度、低对比度、光照不均、文档褶皱和拍照畸变等质量问题的影响，特别对图像二值化处理、主体区域定位、单字符切分等模块影响很大，严重制约字符的识别率提升。

（a）图像模糊　　　　　　　　　　　（b）光斑遮挡

（c）图像过暗　　　　　　　　　　　（d）褶皱

图 12-6 图像质量问题

基于深度学习实现的 OCR 方案具有更高的识别率和更好的鲁棒性（Robust）。特别是 Transformer 模型的引入，其中的 Self-Attention 机制代替 LSTM，不但解决 LSTM 不能并行计算的缺点，而且比 LSTM 支持更长距离的依赖关系的学习，具有比 CNN 更强的特征提取能力，极大提升 OCR 的识别率和鲁棒性。[①]

衡量 OCR 的指标主要包括以下几个方面：

a. 模型准确率。OCR 中模型准确率分为字符识别率和字段识别率，定义如下：

字符识别率 = 识别对的单字符数 / 总的字符数

---

① 更多资料可参考 PaddleOCR：https://github.com/PaddlePaddle/PaddleOCR（《智能文字识别 OCR 能力测评与应用白皮书》），实现 OCR 技术从感知到认知的转变

字段识别率 = 单个字段中全部字符识别对的字段数 / 总的字段数

b. 服务响应时间。服务响应时间 = 模型推理时间 + 网络传输的时间。

c. 服务处理能力，包括 QPS（Query Per Second，系统每秒响应的次数）。

12.2 OCR 模块

基于深度学习的 OCR 中的模块包括：

a. 版式分析（如基于 YOLOv8 模型实现的识别主体区域检测）；

b. 文本定位（以 YOLOv8、CTPN、PSENet、Pixel link、DBNet 等为代表）；

c. 字符识别（以 CRNN OCR、Attention OCR 等为代表）；

d. OCR 结构化（通用结构化模型、NLP、经验知识规则、空间拓扑关系、大模型）；等等。

12.2.1 识别主体区域检测

识别主体区域检测是实现图像中待识别的主体区域的定位。如文档识别的版式分析中，需要先对其中的图片元素进行检测，这个图片元素便是待识别的主体区域。又比如发票 OCR 中先将购物小票区域检测出来，然后做下一步的字符定位，这里的发票区域就是识别主体区域。这样的好处是减少下一步字符定位的计算量。我们可以采用 YOLOv8 模型实现这个待识别主体区域检测，如图 12-7 所示。

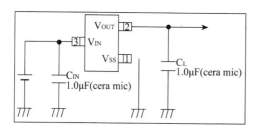

图 12-7　YOLOv8 检测文档中图片元素 / 发票主体区域[①]

12.2.2 文本定位

文本定位算法主要分为基于锚框（Anchor-based）回归和像素（Pixel-based）分割两类。前者以 CTPN、Faster R-CNN、SSD、RFBNet、YOLOVx、SegLink、TextBoxes、

---

① 左图来源：XC6206 芯片手册

TextBoxes++ 等模型为代表，后者主要以 EAST、PSENet、DBNet[①] 等模型为代表。

再简单介绍一下 DBNet 模型。它是基于像素分割的文本定位算法，创新引入可微分二值化模块（Differentiable Binarization，DB），实现每一个像素点的自适应二值化（不同于传统的固定阈值二值化算法），其中的二值化阈值是通过模型训练得到的，如图 12-8 所示。

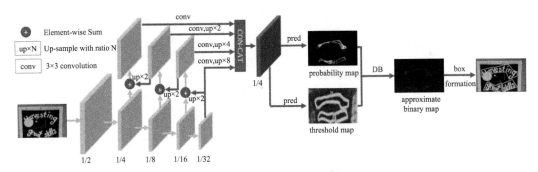

图 12-8　DBNet 模型架构[②]

DBNet 文本定位效果如图 12-9 所示：

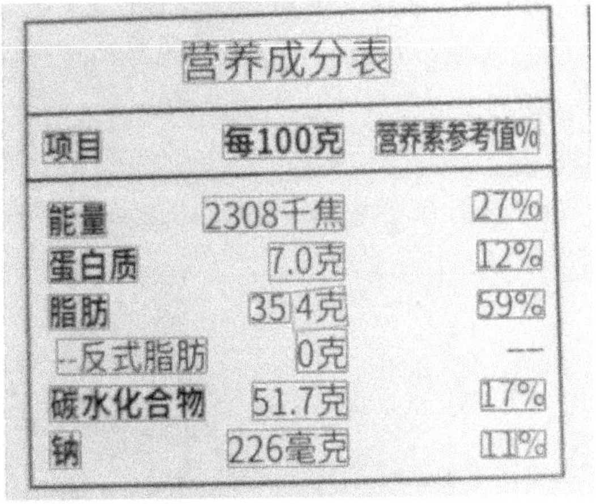

图 12-9　DBNet 文本定位效果

12.2.3　字符识别

字符识别的模型有 CRNN、Attention OCR、Transformer OCR 等。在这里重点讨论 CRNN 模型。CRNN 模型网络包括卷积层、循环层、转录层三部分，如图 12-10 所示。

---

① 代码仓库：https://github.com/MhLiao/DB

② 图来源：Liao M, Wan Z, Yao C, et al.Real-Time Scene Text Detection with Differentiable Binarization. //2020: 11474-11481.DOI:10.1609/aaai.v34i07.6812

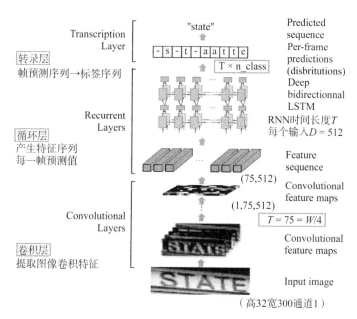

图 12-10 CRNN 网络结构[1]

其中的卷积层实现对图像的卷积特征提取[经过 cv2.resize 得到 $32 \times W \times 1$ 大小，通过卷积计算得到 $1 \times (W/4+1) \times 512$ 大小卷积特征]，循环层实现序列特征的提取（通过 BLSTM 模块处理，然后经过 Softmax 处理，得到的列向量中每个元素代表对应的包含空格的字符预测概率），转录层（CTC loss）实现推理序列去冗余并合并得到一个预测序列。

12.3 隐私 OCR

以上部分我们已经讨论了 OCR 的基本模块，接下来重点讨论隐私 OCR。何为隐私 OCR？它是以保护用户隐私安全为前提的 OCR。

12.3.1 隐私 OCR 流程

在日常生活中的个人证件、交易票据、银行卡、商业合同等影像文件中，都含有一些敏感信息，如果需要使用 OCR 识别功能，该如何保护隐私呢？本节介绍一种基于端侧实现的敏感信息智能脱敏方案。图 12-11 为隐私 OCR 的实现流程。隐私 OCR 和其他 OCR 的不同在于以下 3 点：

---

[1] 图来源：Shi B , Bai X , Yao C .An End-to-End Trainable Neural Network for Image-Based Sequence Recognition and Its Application to Scene Text Recognition. IEEE Transactions on Pattern Analysis and Machine Intelligence，2017. DOI: 10.1109/TPAMI.2016.2646371

（1）脱敏模块实现敏感信息的近物理层次智能脱敏。所谓的近物理层次脱敏是指将敏感信息直接像素级别修改，防止上传到云端服务造成不可知风险。

（2）端侧识别和云端识别相结合的方案。端侧只需要去识别像素很小的切片图片中的敏感信息，好处是不要很大的算力、节约功耗和端侧保护敏感信息的安全。云端服务可依托其更加复杂的模型去实现不同的版式分析及更多复杂的算法功能，此外云端的大算力保证识别的响应时间要求。当然，如果待识别的需求很简单，比如只需要识别其中的票据金额、账户名称等，那就直接使用端侧识别，就更好了。

（3）端侧端云结果融合。

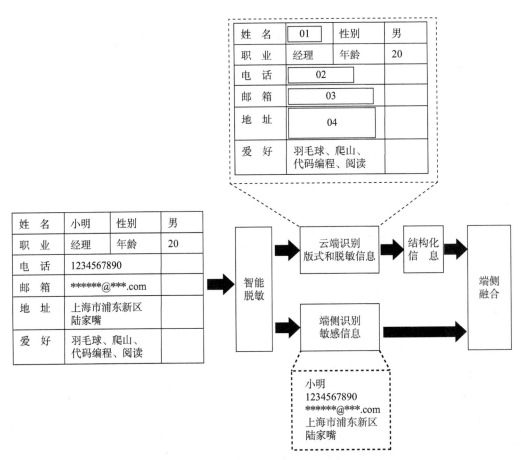

图 12-11　隐私 OCR 实现流程

12.3.2　端侧敏感信息智能脱敏方法

以上的敏感信息的智能脱敏是如何实现的呢？在这里我们介绍一种目前可行的方案。如图 12-12 所示，用手指点击需要脱敏的区域，然后在脱敏的区域通过目标检测的模型（可以使用 YOLOv8 模型）检测候选的文本行框，如图 12-12 中的①②③④所示，用户确定需要定位的文本行即可。

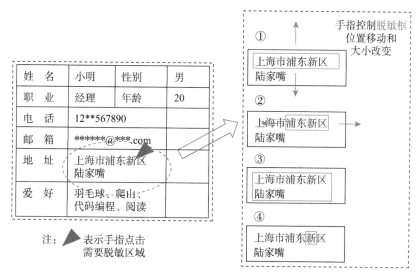

图 12-12　端侧敏感信息智能脱敏

使用同样的算法，我们可以实现手写签名脱敏、印章脱敏等，如图 12-13 所示。

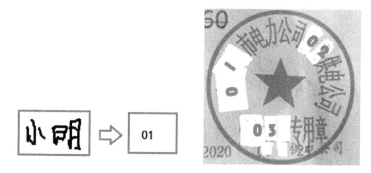

图 12-13　端侧手写签名脱敏和印章脱敏

我们还将继续在端侧将敏感信息的脱敏功能做得更加智能，更加细分化，更加有效用。不仅仅是手机，还有汽车等设备端，为用户使用 AI 服务保驾护航。AI 安全是迫在眉睫的大事，马虎不得，敷衍不得。它是用户体验的基座。它是船，否则我们就是裸泳。我们每多做一点，就是一次很大的进步。

参考文献

[1] Zhao W X, Zhou K, Li J, et al. A survey of large language models [J]. arXiv preprint arXiv:2303.18223, 2023.

[2] OpenAI 官网. Sora 生成 wooly-mammoth 视频 [EB/OL]. [2024-07-06]. https://cdn.openai.com/sora/videos/wooly-mammoth.mp4.

[3] OpenCV 官网. 背景相减方法 [EB/OL]. [2024-07-06]. https://docs.opencab.org/4.x/d1/dc5/tutorial_background_subtraction.html.

[4] Qualcomm 官网. Open, scalable, future-proof ADAS that gets better over time [EB/OL]. [2024-07-06]. https://www.qualcomm.com/products/automotive/automated-driving.

[5] 百度百科. 低空经济 [EB/OL]. [2024-07-06]. https://baike.baidu.com/item/%E4%BD%8E%E7%A9%BA%E7%BB%8F%E6%B5%8E/50884294?fr=ge_ala.

[6] 百度百科. 自动泊车 [EB/OL]. [2024-07-06]. https://baike.baidu.com/item/%E8%87%AA%E5%8A%A8%E6%B3%8A%E8%BD%A6/8187028?fr=ge_ala.

[7] 波士顿动力公司. 波士顿仿人机器人 [EB/OL]. [2024-07-06]. https://www.bostondynamics.com/.

[8] SAR 原理及影像介绍 [EB/OL]. [2024-07-06]. https://www.jl1mall.com/forum/PostDetail?postId=20230627141600331418.

[9] 中国资源卫星应用中心. 雷达影像 [EB/OL]. [2024-07-06]. https://www.cresda.com/zgzywxyyzx/xxgk/kjjx/article/20240425140726594316658.html.

[10] STM32F103xC, STM32F103xD, STM32F103xE Datasheet-production data

[11] EPIC 指令集架构. [2024-07-06]. https://en.wikipedia.org/wiki/Explicitly_parallel_instruction_computing.

[12] Qualcomm Hexagon V66 Programmer's Reference Manual. 80-N2040-42 A. November 17, 2017

[13] https://www.mips.com/

[14] 全志 V853 官方文档. 全志 V853 芯片介绍 [EB/OL]. [2024-07-06]. https://v853.docs.aw-ol.com/.

［15］君正官网. T41-AIPC(普适化)［EB/OL］.［2024-07-06］. http://www.ingenic.com.cn/products-detail/id-2.html.

［16］Arm Developer. Learn the architecture-Introducing the Arm architecture Version 2.1 ［EB/OL］.［2024-07-06］. https://developer.arm.com/documentation/102404/0201/?lang=en, https://developer.arm.com/documentation/102404/0201/Architectre-and-micro-architecture?lang=en.

［17］Shell 函数［EB/OL］.［2024-07-06］. https://www.runoob.com/linux/linux-shell-func.html.

［18］ARM 官网. Arm CPU Architecture: A Foundation for Computing Everywhere［EB/OL］.［2024-07-06］. https://www.arm.com/architecture/cpu.

［19］Arm Developer. Find the Best Processor IP for You［EB/OL］.［2024-07-06］. https://developer.arm.com/ip-products/processors/cortex-a，https://developer.arm.com/ip-products/processors/cortex-r.

［20］意法半导体官网. STM32 系列微控制器［EB/OL］.［2024-07-06］. https://www.st.com/content/st_com/zh.html.

［21］瑞芯微官网. 智能座舱芯片［EB/OL］.［2024-07-06］. https://www.rock-chips.com/a/cn/news/rockchip/2022/1209/1750.html.

［22］Rockchip RV1126 Datasheet

［23］NXP 官网. NXP 的 i.MX 系列产品［EB/OL］.［2024-07-06］. https://www.nxp.com.cn/products/processors-and-microcontrollers/arm-processors/i-mx-applications-processors:IMX_HOME,https://www.nxp.com.cn/products/processors-and-microcontrollers/arm-processors/i-mx-applications-processors/i-mx-6-processors/i-mx-6ull-single-core-processor-with-arm-cortex-a7-core:i.MX6ULL,https://www.nxp.com.cn/products/processors-and-microcontrollers/arm-processors/i-mx-applications-processors/i-mx-6-processors/i-mx-6quad-processors-high-performance-3d-graphics-hd-video-arm-cortex-a9-core:i.MX6Q.

［24］ADI 官网. 降噪耳机［EB/OL］.［2024-07-06］. https://www.analog.com/cn/signals/articles/bowers-wilkins-noise-canceling-headphones.html.

［25］ADI 官网. 处理器和 DSP［EB/OL］.［2024-07-06］. https://www.analog.com/cn/product-category/processors-dsp.html.

［26］全志官网. 全志 R329 芯片［EB/OL］.［2024-07-06］. https://www.aw-ol.com/chips/4.

［27］Qualcomm/Hexagon_SDK/3.5.4/docs/images/80-VB419-108_Hexagon_DSP_User_Guide.pdf

［28］Qualcomm 官网. Snapdragon 8 Series Mobile Platforms［EB/OL］.［2024-07-06］. https://www.qualcomm.com/products/mobile/snapdragon/smartphones/snapdragon-8-

series-mobile-platforms.

［29］Qualcomm Hexagon V66 HVX Programmer's Reference Manual 80-N2040-44 Rev. B October 15, 2018

［30］Qualcomm/Hexagon_SDK/3.5.4/docs/images/80-VB419-108_Hexagon_DSP_User_Guide.pdf

［31］ARM 开发官网．ARM® Mali™ GPU 系列［EB/OL］．［2024-07-06］．https://developer.arm.com/documentation/#&cf［navigationhierarchiesproducts］=%20IP%20Products, Graphics%20and%20Multimedia%20Processors, Mali%20GPUs.

［32］xilinx 官网．XCVU440 FPGA 模块［EB/OL］．［2024-07-06］．https://china.xilinx.com/products/boards-and-kits/1-66ql3z.html, https://www.profpga.com/products/fpga-modules-overview/virtex-ultrascale-based/profpga-xcvu440.

［33］xilinx 官网．AMD Zynq™ 7000 SoCs［EB/OL］．［2024-07-06］．https://china.xilinx.com/products/silicon-devices/soc/zynq-7000.html#productAdvantages.

［34］AMD 官网．Zynq-7000 芯片架构［EB/OL］．［2024-07-06］．https://www.amd.com/zh-cn.html.

［35］全志 v853 文档．NPU 系统介绍［EB/OL］．［2024-07-06］．https://v853.docs.aw-ol.com/npu/dev_npu/.

［36］Rockchip RV1109/RV1126 Technical Reference Manual

［37］IBM 官网．IBM TrueNorth［EB/OL］．［2024-07-06］．https://www.ibm.com/blogs/research/2016/12/the-brains-architecture-efficiency-on-a-chip/.

［38］Intel 官网．神经形态计算 | 超越当今的人工智能［EB/OL］．［2024-07-06］．https://www.intel.cn/content/www/cn/zh/research/neuromorphic-computing.html.

［39］CMake 官网．CMake Tutorial［EB/OL］．［2024-07-06］．https://cmake.org/cmake/help/latest/guide/tutorial/index.html.

［40］The Architecture of Open Source Applications (Volume 1) LLVM［EB/OL］．［2024-07-06］．http://www.aosabook.org/en/llvm.html.

［41］LLVM 源码 等［EB/OL］．［2024-07-06］．https://github.com/llvm, https://llvm.org/, https://llvm.org/docs/LangRef.html, https://llvm.liuxfe.com/.

［42］Qualcomm 开发官网．SNPE［EB/OL］．［2024-07-06］．https://developer.qualcomm.com/sites/default/files/docs/snpe/network_layers.html, https://developer.qualcomm.com/sites/default/files/docs/snpe/snapdragon_npe_runtime.html, https://developer.qualcomm.com/sites/default/files/docs/snpe/aip_runtime.html, https://developer.qualcomm.com/sites/default/files/docs/snpe/overview.html.

［43］CS231n Convolutional Neural Networks for Visual Recognition.［2024-07-06］.

https://cs231n.github.io/convolutional-networks/#overview.

［44］Chellapilla K, Puri S, Simard P. High performance convolutional neural networks for document processing［C］//Tenth international workshop on frontiers in handwriting recognition. Suvisoft, 2006.

［45］ARM 开发官网. NEON/SIMD［EB/OL］.［2024-07-06］. https://developer.arm.com/documentation/den0013/d/Introducing-NEON/SIMD.

［46］Lavin A, Gray S. Fast Algorithms for Convolutional Neural Networks［C］// IEEE. IEEE, 2015.

［47］Shi F, Li H, Gao Y, et al. Sparse Winograd Convolutional neural networks on small-scale systolic arrays［J］. 2018.

［48］Intel 官网. Intel® Intrinsics Guide［EB/OL］.［2024-07-06］. https://www.intel.com/content/www/us/en/docs/intrinsics-guide/index.html.

［49］Nvidia 官网. 端到端自动驾驶汽车开发平台［EB/OL］.［2024-07-06］. https://www.nvidia.cn/deep-learning-ai/solutions/inference-platform/automotive/, https://www.nvidia.cn.

［50］OpenCL 官网. OpenCL is Widely Deployed and Used［EB/OL］.［2024-07-06］. https://www.khronos.org/opencl/.

［51］DBNet 模型代码仓库.［2024-07-06］. https://github.com/MhLiao/DB.

［52］Shell 脚本及应用［EB/OL］.［2024-07-06］. http://staff.ustc.edu.cn/~ccwang/Linux_ch5_1.pdf.

［53］Android 开发官网. Vulkan 开发文档［EB/OL］.［2024-07-06］. https://source.android.google.cn/devices/graphics/implement-vulkan.

［54］Vulkan 开发官网. Vulkan［EB/OL］.［2024-07-06］. http://vulkan.gpuinfo.org/listdevices.php?platform-linux.

［55］Intel 官网. Cache Blocking Techniques［EB/OL］.［2024-07-06］. https://www.intel.cn/content/www/cn/zh/developer/articles/technical/cache-blocking-techniques.html.

［56］Jacob B, Kligys S, Chen B, et al. Quantization and Training of Neural Networks for Efficient Integer-Arithmetic-Only Inference［C］//2017.DOI:10.48550/arXiv.1712.05877.

［57］华为云. 深度学习模型编译技术［EB/OL］.［2024-07-06］. https://bbs.huaweicloud.com/blogs/351263.

［58］TensorFlow 官网. XLA 架构［EB/OL］.［2024-07-06］. https://tensorflow.google.cn/xla/architecture?hl=zh-cn.

［59］https://tvm.apache.org/

［60］OpenCV 开发官网. Geometric Image Transformations［EB/OL］.［2024-07-06］. https://docs.opencv.org/4.x/da/d54/group__imgproc__transform.html#gga5bb5a1fea74ea

38e1a5445ca803ff121ac97d8e4880d8b5d509e96825c7522deb.

［61］https://developer.apple.com/cn/documentation/coreml/

［62］Zeiler M D, Krishnan D, Taylor G W, et al.Deconvolutional networks［C］//Computer Vision & Pattern Recognition.IEEE, 2010.DOI:10.1109/CVPR.2010.5539957.

［63］NEON Programmer's Guide Version: 1.0

［64］Singh P, Verma V K, Rai P, et al. HetConv: Heterogeneous Kernel-Based Convolutions for Deep CNNs［J］. 2019.

［65］https://github.com/irvinxav/Efficient-HetConv-Heterogeneous-Kernel-Based-Convolutio

［66］Guo M H, Lu C Z, Liu Z N, et al. Visual attention network［J］. Computational Visual Media, 2023, 9(4): 733−752.

［67］https://github.com/Visual-Attention-Network/VAN-Classification/blob/main/models/van.py

［68］Luo P, Xiao G, Gao X, et al. LKD-Net: Large kernel convolution network for single image dehazing［C］//2023 IEEE International Conference on Multimedia and Expo (ICME). IEEE, 2023: 1601−1606.

［69］Zhang X, Zhou X, Lin M, et al. Shufflenet: An extremely efficient convolutional neural network for mobile devices［C］.Proceedings of the IEEE conference on computer vision and pattern recognition. 2018: 6848−6856.

［70］Ioffe S, Szegedy C. Batch Normalization: Accelerating Deep Network Training by Reducing Internal Covariate Shift［J］. JMLR.org, 2015.

［71］Srivastava N, Hinton GE, Krizhevsky A, et al. Dropout: a simple way to prevent neural networks from overfitting［J］. The journal of machine learning research, 2014, 15(1): 1929−1958.

［72］Hu J, Shen L, Sun G. Squeeze-and-excitation networks［C］//Proceedings of the IEEE conference on computer vision and pattern recognition. 2018: 7132−7141.

［73］Howard A G. Mobilenets: Efficient convolutional neural networks for mobile vision applications［J］. arXiv preprint arXiv:1704.04861, 2017.

［74］Sandler M, Howard A, Zhu M, et al. Mobilenetv2: Inverted residuals and linear bottlenecks［C］//Proceedings of the IEEE conference on computer vision and pattern recognition. 2018: 4510−4520.

［75］Howard A, Sandler M, Chu G, et al. Searching for MobileNetV3［J］. 2019.DOI: 10.48550/arXiv.1905.02244.

［76］Vaswani A, Shazeer N, Parmar N, et al. Attention Is All You Need［J］. arXiv, 2017.

［77］Han S, Pool J, Tran J, et al. Learning both Weights and Connections for Efficient Neural Networks［J］. MIT Press, 2015.

［78］AutoML 框架［EB/OL］.［2024-07-06］. https://github.com/rhiever/tpot, https://github.com/automl/auto-sklearn,https://autokeras.com/,https://github.com/huawei-noah/vega.

［79］Yao Q, Wang M, Chen Y, et al. Taking Human out of Learning Applications: A Survey on Automated Machine Learning［J］. 2018.

［80］Zller, Marc-André, Huber M F. Benchmark and Survey of Automated Machine Learning Frameworks［J］. 2019.

［81］Gou J, Yu B, Maybank S J, et al. Knowledge Distillation: A Survey［J］. 2020.

［82］知识蒸馏流程［EB/OL］.［2024-07-06］. https://intellabs.github.io/distiller/knowledge_distillation.html.

［83］Horowitz M. 1.1 computing's energy problem (and what we can do about it)［C］//2014 IEEE international solid-state circuits conference digest of technical papers (ISSCC). IEEE, 2014: 10-14.

［84］8-bit Inference with TensorRT［EB/OL］.［2024-07-06］. https://on-demand.gputechconf.com/gtc/2017/presentation/s7310-8-bit-inference-with-tensorrt.pdf.

［85］Shi B, Bai X, Yao C. An end-to-end trainable neural network for image-based sequence recognition and its application to scene text recognition［J］. IEEE transactions on pattern analysis and machine intelligence, 2016, 39(11): 2298-2304.

［86］Nagel M, Fournarakis M, Amjad R A, et al. A white paper on neural network quantization［J］. arXiv preprint arXiv:2106.08295, 2021.

［87］TensorFlow 官网. TensorFlow Lite Hexagon 委托［EB/OL］.［2024-07-06］. https://tensorflow.google.cn/lite/performance/hexagon_delegate?hl=zh-cn.

［88］量化提升策略跨层均衡.［2024-07-06］. https://quic.github.io/aimet-pages/releases/latest/user_guide/post_training_quant_techniques.html#ug-post-training-quantization.

［89］量化提升策略 AdaRound.［2024-07-06］. https://quic.github.io/aimet-pages/releases/latest/user_guide/adaround.html#ug-adaround.

［90］TensorFlow 官网. 量化感知训练综合指南［EB/OL］.［2024-07-06］. https://tensorflow.google.cn/model_optimization/guide/quantization/training_comprehensive_guide?hl=zh_cn.

［91］https://github.com/tensorflow/model-optimization/blob/master/tensorflow_model_optimization/g3doc/guide/quantization/training_example.ipynb

［92］TensorFlow 官网. All symbols in TensorFlow Model Optimization［EB/OL］.［2024-07-06］. https://tensorflow.google.cn/model_optimization/api_docs/python/tfmot/all_symbols.

［93］TensorFlow 官网. 量化感知训练［EB/OL］.［2024-07-06］. https://tensorflow.

google.cn/model_optimization/guide/quantization/training?hl=zh-cn.

[94] AIMET 量化流程.[2024-07-06]. https://quic.github.io/aimet-pages/releases/1.26.0/user_guide/model_quantization.html#aimet-quantization-workflow.

[95] Qualcomm 官网.混合 AI 是 AI 的未来[EB/OL].[2024-07-06]. https://www.qualcomm.cn/.

[96] Machine Learning Platform[EB/OL].[2024-07-06]. https://www.mlplatform.org/.

[97] ARM ComputeLibray 库.[2024-07-06]. https://github.com/ARM-software/ComputeLibrary.

[98] ARM 开发官网. Arm NN SDK[EB/OL].[2024-07-06]. https://www.arm.com/products/silicon-ip-cpu/ethos/arm-nn.

[99] Arm NN 库.[2024-07-06]. https://github.com/ARM-software/armnn.git.

[100] Android 开发官网. Neural Networks API[EB/OL].[2024-07-06]. https://developer.android.google.cn/ndk/guides/neuralnetworks/?hl=zh-cn.

[101] Android 开发官网. NNAPI 运行时[EB/OL].[2024-07-06]. https://source.android.google.cn/docs/core/architecture/modular-system/nnapi?hl=zh-cn.

[102] Android 开发官网. Neural Networks API[EB/OL].[2024-07-06]. https://developer.android.google.cn/ndk/guides/neuralnetworks/?hl=zh-cn.

[103] TNN 深度学习移动端推理框架.[2024-07-06]. https://github.com/Tencent/TNN.git.

[104] OpenVINO 官网. OpenVINO 框架等[EB/OL].[2024-07-06]. https://openvino.org/,https://docs.openvino.ai/latest/openvino_docs_OV_UG_Working_with_devices.html,https://docs.openvinotoolkit.org/latest/omz_models_model_handwritten_simplified_chinese_recognition_0001.html,https://docs.openvino.ai/latest/index.html.

[105] TFLite 开源轻量级深度学习推理框架.[2024-07-06]. https://developers.googleblog.com/2017/11/announcing-tensorflow-lite.html.

[106] Apple 开发官网. Core ML[EB/OL].[2024-07-06]. https://developer.apple.com/documentation/coreml.

[107] Qualcomm 开发官网. Qualcomm Neural Processing SDK for AI[EB/OL].[2024-07-06].https://developer.qualcomm.com/software/qualcomm-neural-processing-sdk, https://developer.qualcomm.com.

[108] Qualcomm 开发官网. Qualcomm AI Engine Direct SDK[EB/OL].[2024-07-06]. https://developer.qualcomm.com/software/qualcomm-ai-engine-direct-sdk.

[109] https://github.com/alibaba/MNN.git

[110] https://google.github.io/mediapipe/solutions/hands

[111] MediaTek 开发官网. NeuroPilot 介绍[EB/OL].[2024-07-06]. https://developer.

mediatek.com/ai/6423c8b6c612745b3a4baa2c.html.

[112] https://docs.opencv.org/4.9.0/index.html

[113] https://developer.arm.com/documentation/102474/latest

[114] https://developer.arm.com/documentation/den0018/latest

[115] CMSIS.［2024-07-06］. https://arm-software.github.io/CMSIS_5/General/html/index.html, https://www.keil.com/pack/doc/CMSIS/NN/html/index.html.

[116] RK3399 主要特性［EB/OL］.［2024-07-06］. https://www.rock-chips.com/a/cn/product/RK33xilie/2016/0419/759.html.

[117] RK3588 主要特性［EB/OL］.［2024-07-06］. https://www.rock-chips.com/a/cn/product/RK35xilie/2022/0926/1656.html.

[118] https://github.com/ARM-software/ComputeLibrary/blob/master/arm_compute/runtime/NEON/NEFunctions.h.

[119] Ne10.［2024-07-06］. http://projectne10.github.io/Ne10/doc/modules.html.

[120] Neon 架构.［2024-07-06］. https://developer.arm.com/Architectures/Neon.

[121] ARM 开发官网. Coding for Neon-Part 5: Rearranging Vectors［EB/OL］.［2024-07-06］. https://community.arm.com/arm-community-blogs/b/architectures-and-processors-blog/posts/coding-for-neon---part-5-rearranging-vectors.

[122] Example-RGB deinterleaving.［2024-07-06］. https://developer.arm.com/documentation/102467/0100/Example---RGB-deinterleaving.

[123] Example-matrix multiplication.［2024-07-06］. https://developer.arm.com/documentation/102467/0201/Example---matrix-multiplication

[124] ARM_NEON_2_x86_SSE.［2024-07-06］. https://github.com/intel/ARM_NEON_2_x86_SSE.

[125] TensorFlow 官网. TensorFlow Lite 委托［EB/OL］.［2024-07-06］. https://tensorflow.google.cn/lite/performance/delegates?hl=zh-cn.

[126] TensorFlow 官网. TensorFlow 实现自定义委托［EB/OL］.［2024-07-06］. https://tensorflow.google.cn/lite/performance/implementing_delegate?hl=zh-cn.

[127] Ruy 矩阵乘法库［EB/OL］.［2024-07-06］. https://github.com/google/ruy.git.

[128] Eigen［EB/OL］.［2024-07-06］. http://eigen.tuxfamily.org/.

[129] Eigen 线性代数运算高性能 C++ 模板库［EB/OL］.［2024-07-06］. https://gitlab.com/libeigen/eigen.git

[130] XNNPACK 库.［2024-07-06］. https://github.com/google/XNNPACK.git.

[131] ARM Cross-compiler［EB/OL］.［2024-07-06］. https://learn.arm.com/install-guides/gcc/cross/.

[132] EUVP dataset [EB/OL]. [2024-07-06]. https://irvlab.cs.umn.edu/resources/euvp-dataset.

[133] Bozic V, Dordevic D, Coppola D, et al. Rethinking Attention: Exploring Shallow Feed-Forward Neural Networks as an Alternative to Attention Layers in Transformers [J]. arXiv preprint arXiv:2311.10642, 2023.

[134] Fan Q, Huang H, Guan J, et al. Rethinking local perception in lightweight vision transformer [J]. arXiv preprint arXiv:2303.17803, 2023.

[135] Hong J, Fulton M S, Sattar J. TrashCan 1.0 An Instance-Segmentation Labeled Dataset of Trash Observations [J]. 2020.

[136] Santoro G, Turvani G, Graziano M. New logic-in-memory paradigms: An architectural and technological perspective [J]. Micromachines, 2019, 10(6): 368.

[137] NXP官网. DSP56F826 Block Diagram [EB/OL]. [2024-07-06]. https://www.nxp.com.cn/products/processors-and-microcontrollers/additional-mpu-mcus-architectures/digital-signal-controllers/16-bit-56800-dsc-core/digital-signal-controller:DSP56F826.

[138] Nvidia官网. NVIDIA Jetson TX2 [EB/OL]. [2024-07-06]. https://www.nvidia.cn/autonomous-machines/embedded-systems/jetson-tx2/.

[139] Nvidia官网. NVIDIA Jetson 用于下一代机器人 [EB/OL]. [2024-07-06]. https://www.nvidia.cn/autonomous-machines/embedded-systems/.

[140] STM32 ARM Cortex 32位微控制器 [EB/OL]. [2024-07-06]. https://www.st.com/zh/microcontrollers-microprocessors/stm32-32-bit-arm-cortex-mcus.html.

[141] Shi W, Caballero J, Huszár F, et al. Real-time single image and video super-resolution using an efficient sub-pixel convolutional neural network [C] //Proceedings of the IEEE conference on computer vision and pattern recognition. 2016: 1874-1883.

[143] Options That Control Optimization [EB/OL]. [2024-07-06]. https://gcc.gnu.org/onlinedocs/gcc/Optimize-Options.html.

[144] MediaPipe部署框架. [2024-07-06]. https://www.mediapipe.dev/,https://developers.google.cn/mediapipe/solutions.

[145] MediaPipe项目代码. [2024-07-06]. https://github.com/google/mediapipe

[146] A set of free tools to enable Edge AI on STM32 MCU and MPUs [EB/OL]. [2024-07-06]. https://www.st.com/content/st_com/zh/ecosystems/artificial-intelligence-ecosystem-stm32.html?icmp=tt17409_gl_pron_sep2020#stann-ai-overview.

[147] Intel Advances Neuromorphic with Loihi 2, New Lava Software Framework and New Partners [EB/OL]. [2024-07-06]. https://www.intel.com/content/www/us/en/newsroom/news/intel-unveils-neuromorphic-loihi-2-lava-software.html.

[148] Debugging with gdb. https://sourceware.org/gdb/current/onlinedocs/gdb.pdf

[149] Desktop GPUs and CPUs Performance Ranking [EB/OL]. [2024-07-06]. https://ai-benchmark.com/ranking_IoT.html,https://ai-benchmark.com/ranking.html.

[150] 方辉云, 何苗, 陈琛, 等. 计算机组成原理 [M]. 武汉: 华中科技大学出版社, 2016.

[151] Intel® 64 and IA-32 Architectures Software Developer Manuals [EB/OL]. [2024-07-06]. https://www.intel.cn/content/www/cn/zh/developer/articles/technical/intel-sdm.html.

[152] Neon [EB/OL]. [2024-07-06]. https://developer.arm.com/Architectures/Neon#Technical-Information.

[153] NEON Programmers Guide. [2024-07-06]. https://static.docs.arm.com/den0018/a/DEN0018A_neon_programmers_guide_en.pdf.

[154] ARM NEON Intrinsics. [2024-07-06]. https://developer.arm.com/architectures/instruction-sets/simd-isas/neon/intrinsics.

[155] Neon Intrinsics-Getting Started on Android Version 1.0. [2024-07-06]. https://developer.arm.com/documentation/102197/latest/.

[156] ARM NEON 寄存器介绍 [EB/OL]. [2024-07-06]. https://goodcommand.readthedocs.io/zh-cn/latest/knowleadge/arm_neon.h.html.

[157] Neon Intrinsics-Getting Started on Android Version 1.0 [EB/OL]. [2024-07-06]. https://developer.arm.com/documentation/102197/0100.

[160] Lin M, Chen Q, Yan S. Network In Network.2013 [2024-06-20].DOI:10.48550/arXiv.1312.4400.

[161] Liao M, Wan Z, Yao C, et al. Real-time scene text detection with differentiable binarization [C] //Proceedings of the AAAI conference on artificial intelligence. 2020, 34(07): 11474-11481.

[162] Vaswani A. Attention is all you need [J]. Advances in Neural Information Processing Systems, 2017.

后　记

对于工程师，技术的广度是重要的，但技术的深度更关键和更具有竞争力。

技术的深度体现在对某些领域中的一块甚至一个点有深入的研究和经验，特别是本书中提到的量产经验，能够提出和实施合理完善的方案，具备行业前瞻性。

技术和产品的关系应该是融合互进的。产品不是以追求极致的技术为目的，而是去"倾听用户声音"和不断提升用户体验感去不断创新和迭代。在这个融合互进的过程中，技术也在不断创新和发展。

照片拍摄于美丽的枸杞岛。记得那天差不多凌晨 3 点到 4 点的样子，起床，出发！可见路上三三两两的手电筒。大约 40 分钟后，我们到达海岛边一个很大的石壁上，等待着日出。没过多久，看见东方的海平面上越来越红，此处省略一篇小作文。

我们应该多去深度思考，提升场景理解力[①]，做闪闪发光的工程师，开发出更多好的 AI 产品。

---

① 场景理解力是能够对场景有深入理解，去深度思考产品和设计产品，不断迭代产品，提升用户体验。场景理解力也是能够把技术转化成产品的能力。笔者认为场景理解力和数据、算法、算力一起构成 AI 的四要素。

在这里，附上一句笔者的座右铭，互勉。

行胜于言！Actions speak louder than words.

在这里，祝愿每个人都能通过不懈的努力，梦想成真！

葳 葳

2024 年 7 月